# Thermodynamic Analysis for Industrial Refrigeration Systems

# Synthesis Lectures on Mechanical Engineering Series

Synthesis Lectures on Mechanical Engineering series publishes 60–150 page publications pertaining to this diverse discipline of mechanical engineering. The series presents Lectures written for an audience of researchers, industry engineers, undergraduate and graduate students.

Additional Synthesis series will be developed covering key areas within mechanical engineering.

Introduction to Deep Learning for Engineers: Using Python on Google Cloud Platform
Tariq M. Arif
2020

Towards Analytical Chaotic Evolutions in Brusselators
Albert C.J. Luo and Siyu Guo
2020

Modeling and Simulation of Nanofluid Flow Problems
Snehashi Chakraverty and Uddhaba Biswal
2020

Modeling and Simulation of Mechatronic Systems using Simscape
Shuvra Das
2020

Automatic Flight Control Systems
Mohammad Sadraey
2020

Bifurcation Dynamics of a Damped Parametric Pendulum
Yu Guo and Albert C.J. Luo
2019

Reliability-Based Mechanical Design, Volume 2: Component under Cyclic Load and Dimension Design with Required Reliability
Xiaobin Le
2019

Reliability-Based Mechanical Design, Volume 1: Component under Static Load
Xiaobin Le
2019

Solving Practical Engineering Mechanics Problems: Advanced Kinetics
Sayavur I. Bakhtiyarov
2019

Natural Corrosion Inhibitors
Shima Ghanavati Nasab, Mehdi Javaheran Yazd, Abolfazl Semnani, Homa Kahkesh, Navid Rabiee, Mohammad Rabiee, and Mojtaba Bagherzadeh
2019

Fractional Calculus with its Applications in Engineering and Technology
Yi Yang and Haiyan Henry Zhang
2019

Thermodynamic Analysis for Industrial Refrigeration Systems
John F. Gardner, Ph.D., P.E.

ISBN: 978-3-031-79704-0     paperback
ISBN: 978-3-031-79705-7     ebook
ISBN: 978-3-031-79706-4     hardcover

DOI 10.1007/978-3-031-79705-7

A Publication in the Springer series
*SYNTHESIS LECTURES ON MECHANICAL ENGINEERING SERIES*

Lecture #33
Series ISSN
Print 2573-3168    Electronic 2573-3176

# Thermodynamic Analysis for Industrial Refrigeration Systems

John F. Gardner, Ph.D., P.E.
Boise State University

*SYNTHESIS LECTURES ON MECHANICAL ENGINEERING SERIES #33*

## ABSTRACT

The vapor compression cycle (VCC) underpins the vast majority of refrigeration systems throughout the world. Most undergraduate thermodynamics courses cover the VCC, albeit in a cursory fashion. This book is designed to offer an in-depth look at the analysis, design and operation of large-scale industrial ammonia-based refrigeration systems.

An important feature of this work is a treatment of computer-aided analysis using Cool-Prop, an open source resource for evaluating thermodynamic properties. CoolProp can be incorporated into a large number of common computational platforms including Microsfot Excel, Python, and Matlab, all of which are covered in this book.

## KEYWORDS

thermodynamics, refrigeration, ammonia refrigeration, multi-stage refrigeration cycles, industrial refrigeration systems

# Contents

# Preface

In 2012, I became the director of one of the roughly two dozen Industrial Assessment Centers (IAC) located in colleges of engineering across the United States. Funded by the U.S. Department of Energy, the IAC program has two main goals: (1) to offer assistance to small and medium manufacturing enterprises by helping them save resources and money and (2) to educate and train the next generation of industrial energy efficiency engineers. As it turned out, this program also provides a hard-won education for the faculty members who are engaged in this program and I found myself on a steep learning curve indeed.

Much of what is required for industrial energy efficiency is based on topics that are well covered in a standard mechanical engineering program, but a few topics required deeper study. Of those, I found industrial refrigeration to be one that had scant resources designed for, and accessible to, the typical upper division undergraduate student. At one end of the spectrum, classical thermodynamics texts [Mor14] cover the vapor compression cycle but not the details of the industrial implementation of such, particular with regard to multi-stage refrigeration. On the other end, we find copious trade publication aimed at the practitioner but with little coverage of the underlying thermodynamics, the best of which is exemplified by the excellent monograph by the good people and Cascade Engineering [WMB04].

This offering is an attempt to fill that void with a text that is accessible to anyone with the first thermodynamics course under their belt, but will give them enough knowledge and experience so that they can navigate the engine room of an industrial facility, perform a thermodynamic analysis of the system and offer some suggestions for operational improvement or equipment modification that can increase the efficiency and efficacy of the system.

John F. Gardner, Ph.D., P.E.
May 2021

# Acknowledgments

I am pleased to acknowledge the US Department of Energy's Industrial Assessment Center program for the inspiration that led to this book. This long-lived program (which actually pre-dates the inception of he Department of Energy itself) has impacted literally thousands of engineers and their professors since its beginning in 1976 (as the Energy Analysis and Diagnostic Centers) and continues to do so to this day. To the program manger at the Advanced Manufacturing Office at DOE, John Smegal and the IAC Field Manger at Rutgers University, Mike Muller (V 1.0), I owe a special thanks. In addition, I would like to acknowledge the assistance of Dr. Robert Comparin at Emerson, Tom Simenc at Cascade Energy and Brian Emtmen at Boise State for their assistance in obtaining photographs of the equipment in Chapter 1. I also thank the students whom I have had the privilege to guide through the challenges of the 'real world' as part of the Boise State/University of Idaho Industrial Assessment Center who helped me hone this message in a way that is, I hope, most impactful to the educational process. Finally I would like to express my gratitude to the wonderful and hard-working colleagues at Boise State University whose dedication to student learning serve as a lasting inspiration to all of us who toil in the trenches of the university classroom.

John F. Gardner, Ph.D., P.E.
May 2021

CHAPTER 1

# Introduction to Industrial Refrigeration

## 1.1 INTRODUCTION

There are few technologies of the modern world that are as wide-spread or have influenced quality of life on this planet as much as refrigeration. The development of the vapor compression cycle (VCC) goes back to the first half of the 19th century and can be found nearly every place humanity occupies. It has revolutionized our food supply system and the design of our residences. The devices built on the thermodynamic principles of the VCC are so reliable that we are often not even aware of their very existence, let alone the details of their operation.

VCC systems can span the gamut from small systems found in the mini-fridges common in college residence halls to huge behemoths required for a warehouse to store the output of a food processing plant. Yet the underlying principles in these two examples are the same.

But as we scale the VCC up to industrial scale, we find some differences worthy of their own treatment. Specifically, the appearance and operation of the hardware are significantly different. Industrial systems tend to utilize a different range of refrigerants and many utilize two stage of compression to maximize efficiency and efficacy of the overall system.

In this section, we exam the typical hardware of large-scale refrigeration systems and acquaint the reader with the function and appearance of the major components one may find in them.

## 1.2 THE COMPONENTS OF INDUSTRIAL REFRIGERATION

As we will describe in detail in Chapter 3, all refrigeration cycles based on the vapor compression cycle have four basic components through which the working fluid (the refrigerant) flows in a cyclical fashion: an evaporator where the heat is removed from the space being cooled; the compressor which takes the low-pressure vapor from the evaporator and raises the pressure (and temperature) of the vapor; the condenser which takes the high-pressure vapor from the compressor and rejects some of its heat to the environment, (thus condensing the vapor to a liquid) and an expansion device (often a simple valve or even a capillary tube) that meters the condensed liquid from the high-pressure condenser to the low-pressure evaporator. The ther-

Figure 1.1: Typical industrial evaporator unit (Tom Simenc/Cascade Energy).

modynamic analysis of these devices will be presented in subsequent chapters, but for now we will demonstrate the typical physical appearance of these devices.

## 1.2.1    EVAPORATORS

The evaporators are the "business end" of the refrigeration system. You will find them in the space being chilled. For large spaces such as warehouses, they are suspended from the ceiling or high on the walls and they appear as finned heat exchangers with integrally mounted fans. The fans direct air from the cools space across the heat exchanger where it gives up some of its thermal energy to the refrigerant in the evaporator. Inside the heat exchanger, the refrigerant, which has just transitioned to lower pressure through the expansion valve, enters in the saturated (both liquid and vapor) state. As it absorbs the heat from the surrounding it steadily transitions to higher quality, that is, toward 100% vapor. Figure 1.1 shows a typical evaporator.

Two things are worth noting about evaporators. First, the humidity in the space tends to condense on the evaporator fins and must be drained away. In spaces below the freezing point, the condensate freezes on the fins, creating a frost coating. Over time, the frost coating impedes the thermal transfer and must be removed through some form of defrosting. Second, the efficiency of the fan motor takes on a higher importance for evaporators. Inefficient motors not only waste energy, but that waste energy emerges into the surrounding space as added heat, which adds to the refrigeration load for the system. In recent years, effort has gone into the development of high efficiency fan motors to address this issue. Currently, electronically commutated motors (ECM) are recommended for these applications.

Figure 1.2: Typical hermetic scroll compressor (Emerson Climate Technologies, used with permission).

### 1.2.2    COMPRESSORS

Refrigeration compressors come in various forms, but they generally fall into three broad categories: hermetic, reciprocating, and screw. Other types, such as centrifugal and scroll, can be found but are less common in industrial settings. In the following sections, each of these types are introduced and their various methods of cooling, capacity control, and oil separation are described.

**Hermetic Compressors**

This is a category that encompasses a variety of sub-classes but are similar in appearance in that they are contained in a welded vessel that contains the compressor, an oil bath, the electric motor, and various mechanisms to separate the oil from the compressed vapor. These are typically small systems suitable for units with capacity from fractions of tons to 10–20 tons, as you might find in rooftop conditioning units. The compressor technology itself may be reciprocal, diaphragm, or scroll, but they are all similar in appearance. They are generally cylindrical and nearly always black. Figure 1.2 shows a typical model. There is usually no capacity control

Figure 1.3: Typical screw compressor. Evaporated low-pressure vapor enters the top (white) and leaves the oil separator to the right (pink). The small cylinder in the foreground is the thermosiphon oil cooler (Tom Simenc, Cascade Energy).

associated with this class of compressors other than on-off control, as is typical in residential applications. Oil separation is handled with the unit itself often supplemented by a small filter or separator immediately downstream of the compressor. Similarly, heat dissipation is usually passive, allowing the heat to radiate and convect off the vessel surface to the surrounding air.

**Screw Compressors**

Larger refrigeration systems, such as you'll find in food processing or warehouse applications, use larger systems that are not suitable to be packaged in a sealed unit. The screw compressor has become the technology of choice for systems with capacities above 100 tons or so. These systems are identified by noting that the compressor is mounted above a long cylinder (which is the oil separator) and is nearly always driven by a motor with a direct drive connection, as seen in Figure 1.3. Unlike hermetic compressors that cycle on and off in response to various needs, a screw compressor is designed to run continuously and capacity control is achieved in one of two ways. These compressors are fitted with a sliding inlet valve which allows the vapor to enter the screws at various stages of compression. The later the entrance, the less capacity. This approach requires little additional equipment but has the disadvantage of being relatively inefficient at partial load situation, thus being more costly to operate if the system operates at partial load a significant portion of the time. A more efficient approach is to control the speed of the compressor using a VFD (variable frequency drive) on the electric motor. This maintains peak efficiency from 100% capacity down to about 20% capacity. Screw compressors require an oil bath on the screws and therefore the compressed vapor must be allowed to separate out from the oil before moving to the condenser. This is accomplished in the oil separator, which is the long cylinder that makes up the bulk of the unit. The heat of compression can be dissipated several ways including a separate oil cooling system, passive heat transfer from the oil separator, or a thermosiphon system.

### Reciprocating Compressors

In very large systems requiring very cold temperatures, systems with two stages of compression are often used (as described in Chapter 4). In such cases, the lower pressure stage (called the booster stage) is often performed with reciprocal compressors, while the upper stage uses screw compressors. Reciprocal compressors are kinematically similar to an automotive internal combustion engine, with multiple pistons operating from a single crankshaft. Like automotive engines they are typically cooled by means of a water jacket and a separate system to reject the heat of compression. Figure 1.4 shows a typical reciprocating model. Capacity control in reciprocating compressors is achieved through valve unloading. At full capacity, all the inlet and outlet valves in the cylinder heads operate normally. When less capacity is needed, the valves of one or more cylinder are disconnected from the drive train, thus unloading those cylinders and taking them out of the compression process. This approach allows for very efficient part-load operation but allows for only discrete steps in load capacity, dictated by the number of cylinders in the unit.

### 1.2.3    CONDENSERS

Once the heat is removed from the refrigerated space, additional heat is added to the refrigerant through the compression process. The sum of those two quantities of heat must then be rejected to the environment. That role falls to the condenser unit. In this section, we'll introduce two common methods of heat rejection, air cooling and evaporative cooling. An air cooled condenser is commonly used in residential and small refrigeration systems. The cylindrical or box-like appliance resting on a pad outside many single family homes is an air cooled condenser. The hot high-pressure refrigerant from the compressor passes through a network of tubes which have fins attached to them. A fan blows ambient air across those fins and the resultant convective heat transfer allows heat to move from the high temperature refrigerant to the lower temperature air. As long as the saturation temperature of the refrigerant at that pressure is higher than the ambient temperature, the refrigerant cools and eventually condenses. While these are most common for small-scale applications, there are applications where they may be appropriate at large scale, for example in places where water is scarce or the ambient humidity is consistently

Figure 1.4: Typical reciprocating compressor. This is a direct drive compressor with the motor out of the photograph to the right. The refrigeration lines are marked in yellow (©User:Endora6398/ Wikimedia Commons/CC-BY-SA-4.0).

very high. In those cases large commercial condensers maybe be the best choice, as shown in Figure 1.5.

Air-cooled condensers have the advantage of having a straightforward design, require very little maintenance, and consume no water. The downsides are that they require significant power to create the convective air flow and they cannot cool the refrigerant below the ambient air temperature. In fact, the lowest you can expect the refrigerant at the outlet of the condenser would be about 8–12°F (4.5–6.7°C) above ambient.

For cases in which air-cooled condensers are not adequate, a more sophisticated and energy efficient solution is employed in which air is introduced below the condenser heat exchanger and forced upward while at the same time, water is sprayed downward from above. The resulting current forces some of the water to evaporate thus absorbing thermal energy and enhancing the convection from the refrigerant tubes to the environment. The refrigerant in the condenser gives up its heat to the convective air/water mixture and eventually condenses. Hence, these units are called evaporative condensers. The water that is not evaporated falls to the bottom of the unit where it is recirculated to the top. Additional water is introduced to make up for the evaporate. This is demonstrated in Figure 1.6.

Figure 1.5: Air-cooled condenser unit.

Unlike air-cooled condensers, evaporative condensers can drive the temperature outside the refrigerant tubes to a point near the wet bulb temperature of the ambient air. Depending on the relative humidity, that can be 20°F (11°C) or more or more below the nominal dry bulb temperatures (at 20–30% relative humidity). Although evaporative condensers are most effective in dry climates, even at 80% relative humidity the wet bulb temperature can be as much as 5°F (2.8°C) below ambient.

Evaporative condensers are typically very large units mounted on the rooftop or near the building being cooled. Figure 1.7 shows a typical evaporative condenser.

## 1.2.4    CHILLERS AND COOLING TOWERS

While chillers are not a focus of this book, we include a brief discussion of chillers here for two reasons. First, they are ubiquitous in commercial building HVAC applications and second, they physically bear some resemblance to scroll and screw compressors, so it's worth discussing them if for no other reason then to be able to tell the difference.

Chillers are integrated systems that include an entire direct expansion (DX) refrigeration cycle (compressor, condenser, expansion device, and evaporator) in a single piece of hardware. The are characterized by a motor-compressor pair, usually in a direct drive configuration and mounted atop two large horizontal cylindrical vessels. The two large cylinders are heat exchangers that contain the condenser and evaporator. Those cylinders will each be connected to a fluid

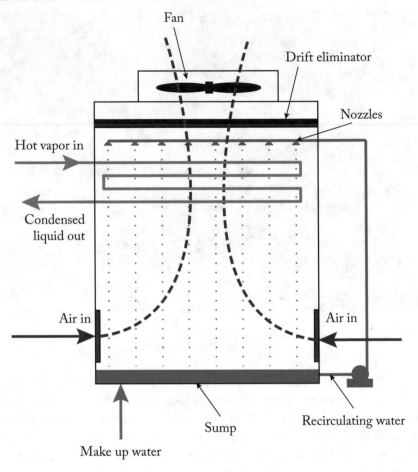

Figure 1.6: Schematic showing the operation of an evaporative condenser.

circuit carrying the working fluid. The condenser will connect to the circuit that pumps water to the rooftop cooling tower (see Section 1.2.5) while the evaporator will connect to a circuit that often carries a water/glycol mixture and used to remove heat from the load. Figure 1.8 shows a typical chiller unit.

As mentioned, the heat from chillers is often rejected to the environment by way of rooftop cooling towers. These are very similar in appearance to evaporative coolers (see Figure 1.7) but are simpler in design and operation. Figure 1.9 shows the schematic operation of a cooling tower. Figure 1.10 shows a photograph of a typical cooling tower.

Chillers have the advantage of being available in many standard sizes and are produced in numbers from a sufficient number of manufacturers to keep capital costs low. You will find them in the machine rooms of large commercial buildings and food retail outlets. Due to the

Figure 1.7: Typical evaporative condenser. Note the refrigerant lines (white) and the water connections (cyan) (Tom Simenc, Cascade Energy).

limited ability of cooling towers to reject heat and the limits of single stage compression, they are generally not suitable for very large-scale loads and for maintaining low temperatures such as a frozen food warehouse.

## 1.2.5    COMPARING COOLING TOWERS AND EVAPORATIVE CONDENSERS

A cursory view of Figures 1.10 and 1.7 might lead one to believe there are no distinguishing characteristics to help tell them apart. However, closer inspection reveals a few significant details that make the distinction easy to spot. The cooling tower is a simpler device that has only two large connections to the rest of the system, the warm water in and the cooler water out. Other connections may be visible to handle the make up water, but they will generally be smaller pipes. On the other hand, the evaporative condenser will nearly always have the water recirculation pipe clearly visible as a vertical pipe rising from a pump which is being fed from the sump. In addition, pairs of refrigeration lines can be seen connected to the upper portion of the unit. If the refrigeration system is ammonia based, depending on local codes and customs, the refrigeration

Figure 1.8: Typical chiller unit. The cylinder in the foreground handles heat rejection and connects to the uninsulated pipes on the left. The rear cylinder contains the evaporator which chills the water carried by the insulated pipes (white) you can see rising behind the motor.

lines are likely to be painted orange and have clearly visible codes indicating what is being carried in the lines.

## 1.3  COMPUTER-ASSISTED THERMODYNAMIC ANALYSIS

One of the most tedious aspects of thermodynamic analysis is the need to look up values of various thermodynamic properties from tables. Undergraduates learn this time-honored practice performing countless homework and exam problems, but the time-consuming nature of property look-up limits the ability to do iterative computations required for design or in-depth analysis. The Engineering Equation Solver[1] is one solution to this challenge and many undergraduate intsitutions have adopted this approach.

In this book, we will look at an open-source package that can add functionality to Microsoft Excel, Python programming environments, and Matlab. This package, called CoolProp, is available as a free download and allows applications to look up the thermodynamic properties of hundreds of fluids and fluid mixtures. In Chapter 2, we walk the use through the down-

---

[1]http://fchartsoftware.com/ees/

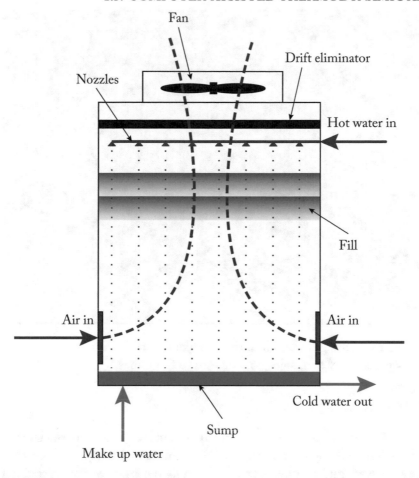

Figure 1.9: Schematic of a typical cooling tower.

load and installation process for CoolProp and illustrate its usage in Excel, Python, and Matlab. Chapter 3 reviews the basic vapor compression cycle, the basis of nearly all modern refrigeration applications, and illustrates the development of an Excel spreadsheet to perform cycle analysis for a residential air conditioner. Further, the "what if" function in Excel is used to show how 2-dimensional parametric analyses can be easily and quickly carried out. Chapter 4 explores four different configurations for two-stage refrigeration systems which are common in industrial settings. Finally, Chapters 5 and 6 present case studies exploring the impact of intercooler pressure and compressor discharge pressure on overall efficiency.

Figure 1.10: Typical cooling tower. Note that, unlike the evaporative condenser in Figure 1.7, this cooling tower has only two major connections to fluid lines.

## 1.4    ABOUT UNITS

In many engineering courses, SI units are normally adhered to as the Imperial Unit system slowly fades away in the engineering community. However, industrial refrigeration systems rarely get much attention in academic coursework and are mainly the realm of the practicing engineering and technical communities. In these settings, the imperial units of BTU, psi, and °F are far more common than kWh, Pascals, and K. To make the results of the analyses presented in this book more accessible to the practicing engineer or technician, Imperial units are widely used, although SI units are sprinkled in as well.

CHAPTER 2

# Computational Tools for Thermodynamic Analysis

## 2.1 INTRODUCTION

For as long as the subject of thermodynamics has been taught in engineering programs, values of the properties of substances were found by looking them up in pre-compiled tables. Indeed, the last scores of pages of nearly all thermodynamics texts were made up of those tables for water/steam, pure gases, propane, ammonia, and some refrigerants. In general, those tables were compiled through a combination of rigorous experimentation and the solution of the thermodynamic equations of state.

More recently, tools have become available to allow for the determination of the properties of a wide range of substances in a computational environment. This approach is typified by the National Institutes of Standards and Technology REFPROP library [Joh13]. These routines, which can be incorporated into a large range of computational platforms, remain the "gold standard" in computer-based thermodynamic properties. Access through these routines requires a license from NIST which can put their use out of the reach of most students and many practitioners. Fortunately, the open source community has developed an alternative that can be accessed free of charge and are adequate for the all but the most demanding analyses. CoolProp is a library of routines written in C++ with "wrappers" that allow interface with a wide variety of platforms, as documented on their site[1] [Bel+14].

For the purposes of this book, we will outline the installation and use for three platforms: Microsoft Excel, Python (Anaconda Spyder), and Matlab.

## 2.2 COOLPROP: EXCEL

The spreadsheet has become the go-to tool for a large portion of engineering analysis and design, so it's not surprising that we should focus on the use of Microsft Excel, the most common spreadsheet application, to perform thermodynamic analyses using CoolProp.

---

[1]http://www.coolprop.org

## 2.2.1   INSTALLING COOLPROP FOR EXCEL

CoolProp is configured as an add-in for Excel and the following is a description from the Cool-Prop website:

> The add-in for Microsoft Excel can also be installed from this package and it auto-matically determines which of the two available add-ins is needed. There is Cool-Prop.xla for Microsoft Office versions prior to version 2007 and CoolProp.xlam for more recent Office installations. The add-ins are installed into the user's add-in di-rectory and they are activated by default. Please note that the Excel add-in includes the installation of some of the shared libraries. Upon installation, an example file is placed on the user's desktop. This file demonstrates some of the features of the Excel wrapper and can be moved or deleted freely, it will not be removed by the uninstaller. Have a look at the dedicated page for more information.[2]

The installer link in CoolProp takes you to a third-party site dedicated to distributing open source software where you need to navigate the file structure to find the binary Win-dows (or Mac) installer. The directory path on the distribution site for the Windows installer is: Home/CoolProp/6.4.1/Installers/Windows.

From there you can download the Windows installer. As of this writing, the Windows installer file name was CoolProp_v6.4.1.0.exe which is the installer for version 6.4.1 of the pack-age. Later versions will be named accordingly.

Download and run this file. The installer may ask you to exit existing programs (e.g., Excel) and then displays the dialog box seen in Figure 2.1.

Make sure you use the correct architecture (most machines are now 64-bit) and the Excel add-in. The installer will then copy the appropriate files into the default directors. Now go to Excel and follow the instructions for "managing add-ins." Unfortunately, the details of this process tend to be dependent on the version of Excel you are using, but the following generic steps from the CoolProp instructions should get you where you need to go.

1. Open Excel.

2. Go to the menu File–>Options–>Add-Ins.

3. At the bottom, select Manage: Excel Add-ins, then click the Go... button.

4. Click the Browse button on the Add-in Manager panel.

5. Browse to the file CoolProp.xlam you downloaded and select it.

6. Make sure the CoolProp Add-in is selected (box checked) and close the Add-in Manager.

---

[2]http://www.coolprop.org

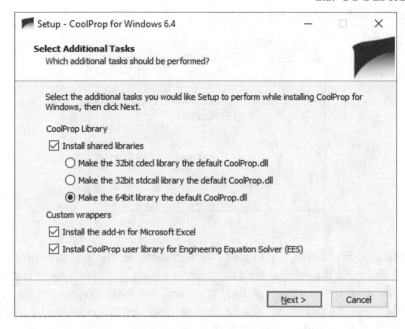

Figure 2.1: Choose the appropriate architecture and applications.

7. Open the file TestExcel.xlsx (now on your desktop) and try to re-evaluate one of the cells; the CoolProp formulas should all be working now. (To recalculate the entire worksheet, press Ctrl-Alt-Shift-F9.)

## 2.2.2    USING COOLPROP IN EXCEL

The fundamental functionality that CoolProp provides is the ability to compute the value of any thermodynamic property once the state is determined. As you will recall, a thermodynamic state of a substance is determined if one knows the value of two independent properties. Following this logic, the functions need to know what property you wish to know and the two independent properties you do know. And it should go without saying that the function needs to know the substance for which the properties are specified. The Excel functions included in the add-in follow this logic. The reader is encouraged to read up on the detailed functionality and the example spreadsheet file (TestExcel.xlsx) shows you the templates for the majority of the calls you'll be needing. For this illustration, we'll focus on the two we're most likely to need for the analysis of refrigeration systems.

In cell D12 in the TestExcel file, you'll find this function call:

```
=PropsSI("T","P",101325,"Q",0,"Water")-273.15
```

Table 2.1: Property abbreviations for PropsSI calls

| Abbreviation | Property | Units |
|:---:|:---|:---:|
| D | Mass density | kg/m$^3$ |
| H | Enthalpy | J/kg |
| P | Pressure | Pa |
| Q | Quality | – |
| S | Entropy | J/kg/K |
| T | Temperature | K |
| U | Internal energy | J/kg |

The function has five arguments, the first of which determines the property that is going to be returned, while the next four are label:value pairs in which you specify any two independent properties that determine the state. The final argument is a text string denoting the substance. In this case, the function call above can be read as: find the temperature of water for which the pressure is 101,325 Pa and the quality is 0. In other words, the function will determine the temperature of pure water at the saturated liquid state at one standard atmosphere of pressure, i.e., the boiling point of water. Note that CoolProp works entirely in SI units so the final part of the equation is simply converting the resulting temperature from K to °C.

The number of properties that can be computed and specified in this manner are extensive, but Table 2.1 lists a few of the standard properties and their abbreviation for the PropsSI call which can be used as either input or output.

In addition to the generic PropsSI function, which computes the properties for pure and pseudo-pure substances (like some refrigerant mixtures), there is also a specialized function for humid air. Cell D36 is TestExcel is an example of that function.

```
=HAPropsSI("D","T",293.15,"P",101325,"R",0.5)
```

There are some differences between the abbreviations used in HAPropsSI and PropsSI as can be seen in the CoolProp documentation, a portion of which is listed in Table 2.2. As was the case with PropsSI, any of these properties can be used to specify either input or output properties, with the exception of pressure. For HAProps, pressure can only be specified as an input property.

For this particular example, the function finds the dew point temperature (D) for humid air at 293.15 K (20°C, 68°F), at a pressure of 101,325 Pa (1 std atmosphere) and a relative humidity of 50%. Note that since humid air is a mixture, we need to specify two independent properties and the ratio of the mixture as well.

Table 2.2: Property abbreviations for HAPropsSI calls

| Abbreviation | Property | Units |
|---|---|---|
| D, DewPoint, Tdp | Dew point temperature | K |
| H | Enthalpy | J/kg dry air |
| P | Pressure | Pa |
| R | Relative humidity | – |
| S | Entropy | J/kg dry air/K |
| T,  Tdb | Dry bulb temperature | K |
| B, WebBulb, Twb | Wet bulb temperature | K |
| W, Omega, HumRat | Humidity ratio | kg water/kg dry air |

### 2.2.3    AN EXAMPLE IN EXCEL

As an example, let's use the CoolProp function to compute the saturation dome of a substance and plot it on both a T-S diagram and a P-H diagram. As we know from thermodynamics, the saturation dome refers to the lines that separate the liquid, mixed, and vapor phases of a substance. The lines merge to a single point at the critical temperature, which is unique to a substance.

Start the exercise by reserving a cell at the top of the spreadsheet for the name of the substance you're working with. CoolProp is capable of computing the properties for dozens of substances,[3] but for this exercise the focus will be on refrigerants.

Directly under that cell, use the call as shown in TextExcel to compute the critical temperature for that substance. Use the previous cell (describing the substance) in the CoolProp call so that you can easily change the substance. Your spreadsheet should look like the screen shot on Figure 2.2. The formula in cell B2 is:

```
=Props1SI(B1,"Tcrit")
```

Now create a column (in column A) for temperatures and start by referencing the critical temperature. Fill out that column by "counting down" from the critical temperature by 5 Kelvin.

Now create five computed columns, saturation pressure, entropy of saturated liquid, entropy of saturated vapor, enthalpy of saturated liquid, and enthalpy of saturated vapor. For example, if the first temperature in our spreadsheet is in A5 and the name of the substance is in B1, then the first cell in saturated pressure columns should have this equation:

[3]http://www.coolprop.org/fluid_properties/PurePseudoPure.html#list-of-fluids

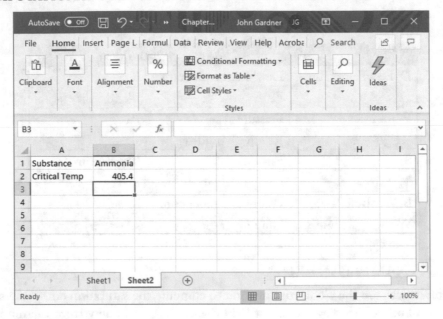

Figure 2.2: Beginning of our spreadsheet.

```
=PropsSI("P","T",A5,"Q",0,$B$1)
```

Similarly, the saturated vapor enthalpy should read:

```
=PropsSI("H","T",A5,"Q",1,$B$1)
```

If you choose the refrigerant R134A, then the top of the spreadsheet should now look like Figure 2.3.

Finally, create two charts using the $x-y$ scatter template. In one chart, plot two data sets with temperature on the $y$-axis and the two entropy columns as the $x$-axis data. That should give you the saturation dome for the T-S diagram, as shown in Figure 2.4. On the second plot, create the saturation dome for the P-H diagram using the pressure and enthalpy columns. The results should look like Figure 2.5.

## 2.3   COOLPROP: PYTHON

As we will see in Chapter 3, the Excel add-in is very powerful for simple analyses, but can be tedious for extensive sensitivity studies or other important design-oriented tasks. For more sophisticated analyses, engineers often find it more efficient to use a programming language

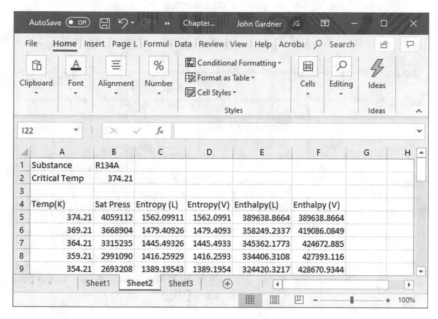

Figure 2.3: Spreadsheet with computed columns.

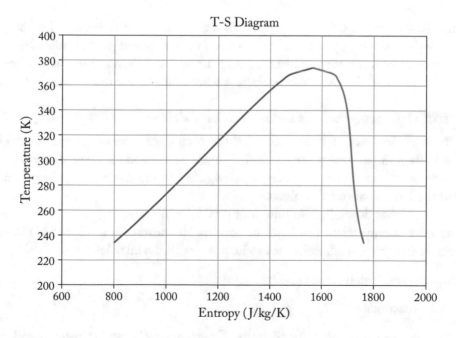

Figure 2.4: Saturation dome on the temperature-entropy diagram for R134A.

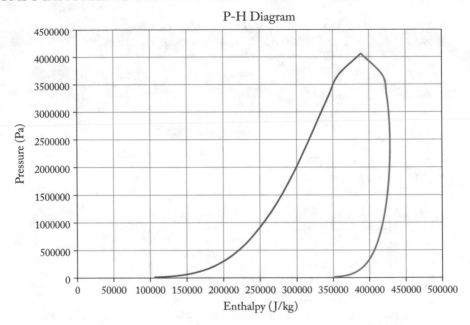

Figure 2.5: Saturation dome on the pressure-enthalpy diagram for R134A.

or scripting environment. CoolProp supports a large number of these, but for the purposes of this text, we'll focus on Python (in the Anaconda environment) and Matlab. In its current implementation, CoolProp supports Matlab only through the Python interface, so installing a Python application and the CoolProp package is a pre-requisite for using CoolProp in Matlab.

## 2.3.1 INSTALLING ANACONDA AND PYTHON

The Anaconda environment is a powerful platform for a large array of computing tasks, and it supports a Python development console called Spyder. Anaconda is open source and free for individuals to download. Start the process by navigating to the Anaconda website[4] and follow the links to the individual free download.

Please note that depending on how your particular Windows environment is configured, you may require administrator privileges to perform this installation. If that is the case, you should contact your system administrator to help you with the installation.

1. Download and install Anaconda.

2. Launch Anaconda.

3. At the navigator (main) screen, click the "Environments" item in the left-hand panel.

[4]https://www.anaconda.com/

4. Click on the green arrow/triangle next to "base (root)."

5. Choose "Open Terminal" from the drop-down menu.

6. In the command prompt terminal that opens up, type: pip install Coolprop.

7. When the operation is complete, close the command window, return to the Anaconda Navigator and click the "home" icon.

8. Launch Spyder from the Anaconda Navigator.

9. In the Console window of Spyder, type these two commands:

   import CoolProp.CoolProp as cp.

   ```
   cp.PropsSI('T','P',101000,'Q',0,'water').
   ```

10. Verify that the result is 373.034 Kelvin.

The call to PropsSI() is essentially the same syntax as the Excel add-in and allows the user to compute thermodynamic properties within the Python programming environment. In addition to this basic functionality, there is a wide range of additional capabilities that can be explored on the CoolProp website.[5]

One of the more interesting capabilities is the combination of Coolprop with the Python plotting package. For example, you can create typical property plots for a given refrigerant. The code in Listing 2.1 shows an example to create TS diagram for the now obsolete refrigerant, R12 (commonly known as Freon).

The resulting diagram is shown in Figure 2.6.

## 2.4    COOLPROP: MATLAB

As discussed previously, you can call CoolProp from Matlab through the Python interpreter. Therefore, the first step in accessing the CoolProp is to install a Python platform and install CoolProp within it, as described in the previous section. Now we need to ensure that Matlab knows where to find the Python interpreter. In the Matlab command window, type:

```
>> pyversion
```

If the response looks like your version of this:

---

[5]http://www.coolprop.org/coolprop/wrappers/Python/index.html#python

**Listing 2.1**   Python code to create a T-S diagram for Freon

```python
import CoolProp
from CoolProp.Plots import PropertyPlot
#
#  Instantiate the 'PropertyPolt' object for Freon in SI units
#
plot = PropertyPlot('R12', 'TS', unit_system='SI', tp_limits='ACHP')
#
# Create 11 lines of constant quality
# if you only want to see the dome, use num=2
#
plot.calc_isolines(CoolProp.iQ, num=11)
#
# Create 10 lines of constant pressure (isobars)
#
plot.calc_isolines(CoolProp.iP, num=10, rounding=True)
#
#  create the plot, modify the iso-lines
#
plot.draw()
plot.isolines.clear()
plot.props[CoolProp.iP]['color'] = 'green'
plot.props[CoolProp.iP]['lw'] = '0.5'
#
#  display and save the plot as a png
#
plot.show()
plot.savefig('FREON-TS.png')
```

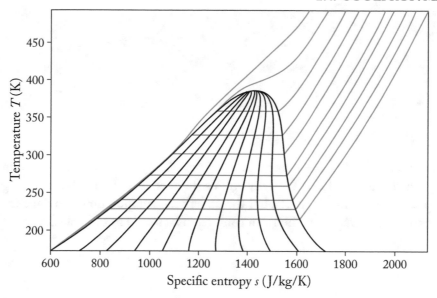

Figure 2.6: T-S diagram for R12 (Freon) with lines of constant pressure and quality.

```
>> pyversion

    version: '3.7'
    executable: 'C:\Users\John Gardner\Anaconda3\python.exe'
    library: 'C:\Users\John Gardner\Anaconda3\python37.dll'
    home: 'C:\Users\John Gardner\Anaconda3'
    isloaded: 0
```

then you are ready to go. If you have installed Python and Matlab, but the pyversion command does not return a Python version, you need to tell Matlab where to find the Python interprester. First find the file py.exe, then issue this command in Matlab:

```
>> pyversion 'c:\windows\py.exe'
```

Make sure you replace the director path with the one that is correct for your machine.
To call CoolProp from the Matlab command line (or a script), use the following syntax:

```
    >> py.CoolProp.CoolProp.PropsSI('T','P',101000,'Q',0,'Water')

ans =

373.0343
```

## 2.5    UNIT CONVERSIONS

The native units for the CoolProp package is strictly SI. However, the refrigeration technical community (in the U.S. at least) typically still uses imperial units. In the following chapters, results are presented in imperial units, but the underlying computations are done in SI. To aid in the conversion, we summarize the main conversion factors here in Table 2.3.

Table 2.3: Common unit conversion factors

| Mass | |
|---|---|
| 1 kg | 0.455 lbm |
| 1 lbm | 2.2 kg |
| **Energy Transfer** | |
| 1 Ton Refrigeration | 12,000 BTU/hr |
| 1 BTU/hr | 8.33E-5 ton refrigeration |
| 1 Ton Refrigeration | 3.517 kW |
| 1 kW | 0.284 ton refrigeration |
| 1 kW | 1.341 hp |
| 1 hp | 0.746 kW |
| 1 kW | 3412 BTU/hr |
| 1 BTU/hr | 2.931E-4 kW |
| **Energy** | |
| 1 kWh | 3412 BTU |
| 1 BTU | 2.931E-4 kWh |
| **Pressure** | |
| 1 psi | 6895 Pa |
| 1 Pa | 1.450E-4 psi |
| 1 bar | 1E5 Pa |
| 1 Pa | 1E-5 bar |
| 1 psi | 0.0689 bar |
| 1 bar | 14.50 psi |
| **Temperature** | |
| $T(°C) = (T(°F) - 32)/1.8$ | |
| $T(°F) = (T(°C) * 1.8) + 32$ | |
| $T(°C) = T(K) - 273.15$ | |
| $T(°F) = T(°R) - 459.67$ | |

CHAPTER 3

# The Vapor Compression Cycle: A Review

## 3.1 INTRODUCTION

The vapor compression cycle applied to refrigeration goes back to the early 19th century and was first used to make ice for breweries and meat packing facilities. The cycle depends on two important characteristics of the liquid/vapor phases of substances. One is that the boiling temperature increases as pressure increases, the other is that mixtures of liquid and vapor phases of a substance (i.e., a saturated mixture) can absorb or reject heat without changing temperature. At low pressure (often sub-atmospheric) the refrigerant can absorb heat by boiling at temperatures below the desired refrigeration target temperature. At higher pressures, the vapor can condense at temperatures far above typical ambient temperatures. Using a compressor to increases the pressure of the evaporated vapor and a valve or orifice to meter the condensed liquid back to the lower pressure state, a cycle can be engineered to continuously pump heat from a low-temperature thermal reservoir to a high-temperature one.

## 3.2 THE CARNOT CYCLE

At approximately the same time various engineers and physicists were attempting to perfect machines to implement the vapor compression cycle, French physicist Sadi Carnot was working on one of the most significant discoveries in thermodynamics and power conversion. In short, he discovered that when converting thermal energy to mechanical energy using a heat engine, there is a theoretical limit to how much of that energy can be converted and that the limit is governed by the temperature of the two thermal reservoirs between which the thermal energy flows. The limit, known as the Carnot Efficiency, can also be applied to the vapor compression cycle, which is essentially a heat engine running in reverse. This theoretical limit can be best visualized by plotting the path of the cycle on the temperature-entropy diagram for the refrigerant. Figure 3.1 shows a schematic of the Carnot refrigeration cycle while Figure 3.2 represents that cycle on a Temperature-Entropy (T-S) diagram.

   The cycle has four processes operating between four states, moving in a counter-clockwise fashion around the diagram and described below. Note that the blue lines in Figure 3.2 are lines of constant pressure (isobars) and the cycle operates between two pressures.

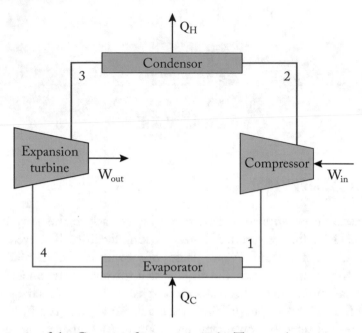

Figure 3.1: Schematic of the Carnot refrigeration cycle. The numbers represent the states of the cycle.

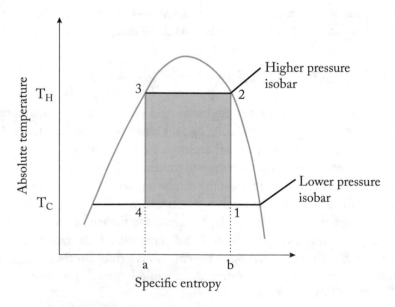

Figure 3.2: Carnot refrigeration cycle on a T-S diagram. The numbers represent the states of the cycle and the flow of refrigerant is counter-clockwise.

- Isentropic Compression: adiabatic and reversible compression from the lower pressure to a higher pressure. State 1 is at the lower pressure and a saturated mixture. State 2 is saturated vapor at the higher pressure. It requires mechanical work to perform this compression.

- Isothermal Condensation: the saturated vapor at state 2 is condensed to state 3, which is saturated liquid. In doing so, the latent heat in state 2 is rejected to the environment.

- Isentropic Expansion: the saturated liquid in state 3 is expanded through an adiabatic and reversible turbine. Useful work is extracted and the resulting state is a saturated mixture at the lower pressure.

- Isothermal Evaporation: the saturated mixture absorbs heat from the surrounding environment causing the some of the liquid to evaporate until it reaches state 1 and the cycle starts anew.

This simple representation allows us to develop some deep insight into the limits of the vapor compression cycle. A complete development of this insight can be found in [Mor14], but for the purposes of this discussion, a simpler approach will suffice.

We start be recalling the relationship ship between entropy, temperature, and heat transfer in Equation (3.1):

$$\delta Q = T dS, \tag{3.1}$$

which states that the heat transfer in a reversible thermodynamic interaction is equal to the temperature of the interaction times the differential change in entropy. For constant temperature interactions (as is the case for the ideal evaporators and condensers), that relationship becomes:

$$\Delta Q = T \Delta S. \tag{3.2}$$

So for the cycle shown in Figure 3.2, the heat transfer at the condenser (which we'll call $Q_H$) and the heat transfer at the evaporator ($Q_C$) are:

$$Q_H = T_H * (S_b - S_a)$$
$$Q_C = T_C * (S_a - S_b). \tag{3.3}$$

The first law of thermodynamics tells us that the energy in must equal the energy out, as shown in Equation (3.4):

$$W_{net} + Q_C = Q_H. \tag{3.4}$$

We now define the Coefficient of Performance for a refrigeration cycle as the ratio of the heat transferred from the cold reservoir (our desired effect) to the net work required of the cycle (the thermodynamic cost):

$$\beta = \frac{Q_C}{W_{net}}. \tag{3.5}$$

Combining Equations (3.3) through (3.5), we find

$$\beta = \frac{T_C * (S_b - S_a)}{T_H * (S_b - S_a) - T_C * (S_b - S_a)}$$

$$\beta = \frac{T_C}{T_H - T_C},$$

(3.6)

where $T_C$ and $T_H$ are expressed in absolute temperature. From this exercise we can draw two conclusions. First, unlike thermodynamic efficiency, the coefficient of performance can be (and usually is) greater than unity. Second, the theoretical limit for the COP is related to the temperatures between which the cycle operates and the closer they are to each other, the higher the COP.

One more thing before we move on to more realistic cycles. It's worth noting that the areas within contours on a T-S diagram represent the specific energy (energy per unit mass of working fluid) of the cycle. Specifically, the area enclosed by the contour drawn through a-4-1-b-a represents the specific thermal energy absorbed by the evaporator while the area defined by a-3-2-b-a represented the specific thermal energy rejected by the condenser. By inference, we conclude that the net work required by the cycle is the area of the rectangle defined by 1-2-3-4-1. Finally, note that by convention, the TS diagram is rarely referenced to absolute zero, so the relative sizes of these rectangles is not often obvious from the figure. For example, we have already stated that the thermal energy absorbed by the evaporator (area of a-4-1-b-a) is generally larger (2–3 times larger) than the net work required (1-2-3-4-1) but that doesn't appear to be the case in Figure 3.2. The reason for this is that the $x$-axis on the plot does not define absolute zero on the temperature scale. If we were to draw the figure to scale, much of the detail of the figure would be lost.

## 3.3   THE IDEAL VAPOR COMPRESSION CYCLE

It can be shown that the Carnot cycle as described in Section 3.2 has the highest performance from an energy perspective. Unfortunately, the Carnot cycle possesses several characteristics that make in impractical for a real system. The most significant issues with the Carnot cycle are the isentropic compression and expansion processes which take place within the saturated region. Compressors work well with vapor and pumps work well with liquids, but a mixture of vapor and liquid is highly problematic from an engineering perspective. Therefore, the compression phase of the cycle is nearly always "pushed" outside the saturate dome and takes place with super-heated vapor. Similarly, isentropic expansion can only occur if work is extracted. While on the face of it, extracting additional work from the system would seem to be attractive, its practicality is limited. Developing a machine that can extract work from a liquid vapor mixture is problematic itself, but in addition, the amount of work that can be extracted is fairly small, typically 10–20% of the amount of work needed for compression. Considering the additional complexity

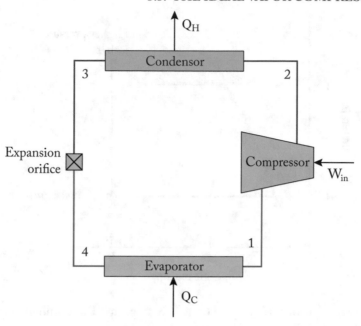

Figure 3.3: Schematic of the ideal vapor compression cycle. The numbers represent the states of the cycle.

and expense required to extract that energy and find a use for it, the turbine is omitted. Instead of isentropic expansion through an adiabatic turbine, the fluid is allowed to expand through an orifice (often a valve of a small tube) in an isenthalpic process. This combination of isentropic compression and isenthalpic expansion is the basis for the Ideal Vapor Compression Cycle with the schematic shown in Figure 3.3 and the resulting T-S diagram shown in Figure 3.4.

As the name implies, the Ideal Vapor Compression Cycle is still not quite an accurate representation of an actual refrigeration system, but it forms a useful basis for engineering analysis. We now list the assumptions incorporated in the ideal cycle and how they differ from real-world systems.

- No pressure drop in the evaporator or compressor: as can be seen in Figure 3.4, the phases of the cycles that represent evaporation and condensation lie along isobars on the T-S diagram. In reality, this could not be the case because some pressure difference is required to ensure that the working fluid moves through the pipes. However, the evaporators and condensers are designed to minimize this pressure drop as it represents wasted energy. Long experience with systems indicates that this is a pretty good assumption.

- State 3 (exit of the condenser) is saturated liquid: it seems unlikely that the actual system would be tuned so that the working fluid is precisely at the saturated liquid state

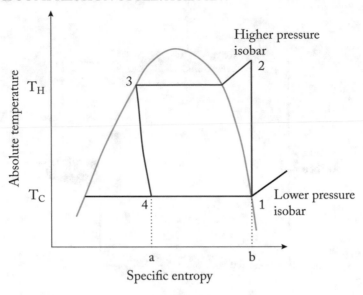

Figure 3.4: Ideal vapor compression cycle on a T-S diagram. The numbers represent the states of the cycle.

as it exits the condenser. In reality, condensers tend to be somewhat over-designed so that the refrigerant is somewhat sub-cooled at the exit. This has limited impact to our analysis because the amount of sub-cooling achieved in these devices is generally quite small and the resulting change in enthalpy is small compared to the change that is achieved through condensation.

- State 1 (exit of the evaporator) is saturated vapor: again, it seems unlikely that a system would preform in such a controlled way. However, it is a good assumption for two reasons. First, for simpler DX (direct expansion) systems, the expansion device is a variable orifice valve controlled to make sure that the exit of the evaporator is a few degrees into the super-heat region. Second, there is often a separator between the evaporator and the compressor to ensure that only vapor is allowed to enter the compressor. The analysis is easily modified to allow for the fact that state 1 is in the super-heat region.

- The compression phase is isentropic: this assumes that the compression is adiabatic, which is highly unlikely. The heat of compression raises the temperature of the compressor which then drives heat transfer to the environment. This can be accounted for if the isentropic efficiency of the compressor is known. The isentropic efficiency allows you to compute a more accurate enthalpy for state 2 once the properties for the state resulting from the isentropic compression is known.

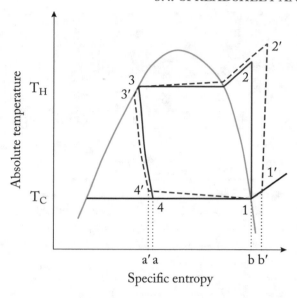

Figure 3.5: Ideal vapor compression cycle on a T-S diagram (solid lines) and a more realistic cycle (dashed lines, states with primes). The numbers represent the states of the cycle.

Figure 3.5 shows what the T-S diagram looks like if these assumption are relaxed. Note that the non-isentropic compression pushes state 2 to a higher pressure and temperature (state 2') while the pressure drops in the condenser and evaporators are reflected in a gradual "glide" in the fluid pressure (and hence, temperature) through the heat exchangers. Finally, note that the entrance of the compressor is to the right of the saturation dome, indicating that the fluid is slightly superheated. Referring to the previous discussion about the interpretations of enclosed areas on the T-S diagram, we note that the real cycle will typically have more heat removal per unit mass working fluid when compared to the ideal cycle (area given by a'-4'-3'-b'-a') but at the cost of more compressor work (area 4'-3'-2'-1'-4'). For the vast majority of applications, the ideal cycle is a fine basis for analysis and these variations are only considered if a great deal is known about the system under consideration.

## 3.4   SPREADSHEET ANALYSIS EXAMPLE

As a way of demonstrating the general analysis techniques used for a VCC and the Excel add-in, we will look at the impact of outside temperature on the operation of a typical residential whole-house air conditioner unit. Below are some performance specifications from a 2-ton Goodman residential HVAC unit utilizing the refrigerant R410a (Model: GSX130241DA).

The first step is to find the saturation temperatures between which the system is operating. In other words, we need to find the temperature of the refrigerant as it is evaporating in the

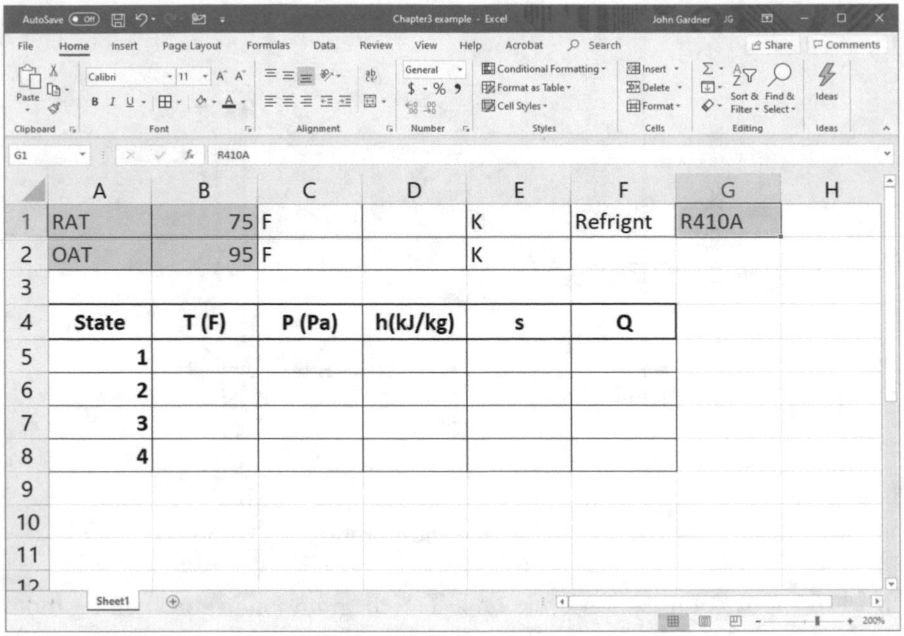

Figure 3.6: Beginning of a spreadsheet to analyze the ideal vapor compression cycle.

indoor evaporator and the temperature of the fluid as it is condensing in the outdoor condenser. While these temperatures are sensitive to a wide range of conditions from indoor humidity, state of charge of the system, filter status, etc.; there are rules of thumb that help us get started at this point.

In general, a well operating residential AC unit will have a condenser saturation temperature that is about 15°F above the ambient Outdoor Air Temperature (OAT). Similarly, the evaporator saturation temperature will be about 30°F below the entering air temperature (that is, the temperature of the air in the home as it is brought back into the air handler to be cooled down). The reason for this should be obvious because the exchange of heat in the condenser and evaporator is driven by the difference in temperature between the refrigerant and the air moving across the exchangers. The $\Delta T$ in the evaporator is considerably larger than that of the condenser due to the fact that the evaporators are generally much smaller than the condensers and hence require a larger temperature difference to drive a similar amount of heat transfer due to the smaller area available for heat exchange.

We start the spreadsheet analysis with these two numbers: Return Air Temperature (RAT) and OAT. Next, we have to build a table that allows us to determine the thermodynamic state of each stage of the VCC cycle (states 1–4 in Figure 3.4). Figure 3.6 shows a screen shot of the beginning of the process. The inputs to this tool are shaded in orange (refrigerant, RAT and

OAT). Note that the sheet expects these temperatures in F but the CoolProp add-in operates in SI units, so we'll add a column to compute those temperatures in K.

Using our rules of thumb and assumptions about the ideal VCC, we can immediately specify the temperatures of states 1 and 4. They are at the saturation temperature of the evaporator, which should be 30°F (16.7°C) below the RAT. Similarly, the temperature of state 3 (saturation temperature of condenser) will be 15°F (8.3°C) above the OAT. Those are easily computed using standard spreadsheet formulas (remember to use the right units when adding/subtracting the ΔTs). Next, recognize that the quality of states 1 and 3 are known and enter those values. Figure 3.7 shows the current state of the spreadsheet.

Since any two independent properties determine the state of a substance, we can now compute all properties for states 1 and 4, by calling on CoolProp (the PropsSI() function) as described in Chapter 2. Next recognize that state 2 is the same pressure as state 3 (they lie on the same isobar in Figure 3.4). Similarly, States 1 and 4 share an isobar, so update the pressure for state 4 accordingly. Figure 3.8 shows the updated spreadsheet. To define states 2 and 4, we still require an additional independent property for each. The assumptions of the Ideal VCC provide those. Recall that it was assumed the compression process is isentropic, which implies that the entropy of state 2 is the same as the entropy at state 1, so you can simply insert an equation in cell E6 to reference E5. Similarly, the expansion process is assumed to be isenthalpic, so the enthalpy of state 4 (cell D8) is the same as the enthalpy of state 3 (cell D7).

Now we can round out the table by inserting the appropriate CoolProp formula in each cell to compute the unknown properties. It is an interesting exercise to complete this table using the traditional table look-up method if for no other reason, it drives home the power of using computational tools to assist in this analysis. In particular, the process of determining the temperature of state 2, which is well into the superheat region, is particularly tedious in that it requires a process of double interpolation. The completed table should now look like Figure 3.9. To help clarify the process, those cells that contain a call to PropsSI() are shaded in green. The cells shaded in orange refer directly to the RAT and OAT inputs (with the rule-of-thumb offsets applied) and the unshaded cells can be inferred from the structure and assumptions of the ideal vapor compression cycle.

To help make the results more accessible and to compare the analysis to the manufacturer's specifications, we'll add an additional state property table with the temperature and pressure in imperial units as shown in Figure 3.10.

Once the specific values for all the state properties are found, we can find the work and heat per unit working fluid for the cycle. Those are easy to compute on the spreadsheet by applying a simple energy balance on each component:

$$
\begin{aligned}
q_{low} &= h_1 - h_4 \\
q_{high} &= h_2 - h_3 \\
w_{comp} &= h_2 - h_1.
\end{aligned}
$$
(3.7)

| | A | B | C | D | E | F | G |
|---|---|---|---|---|---|---|---|
| 1 | RAT | 75 | F | 297.0 | K | Refrignt | R410A |
| 2 | OAT | 95 | F | 308.2 | K | | |
| 3 | | | | | | | |
| 4 | State | T (K) | P (Pa) | h(kJ/kg) | s | Q | |
| 5 | 1 | 280.4 | | | | 1 | |
| 6 | 2 | | | | | | |
| 7 | 3 | 316.5 | | | | 0 | |
| 8 | 4 | 280.4 | | | | | |

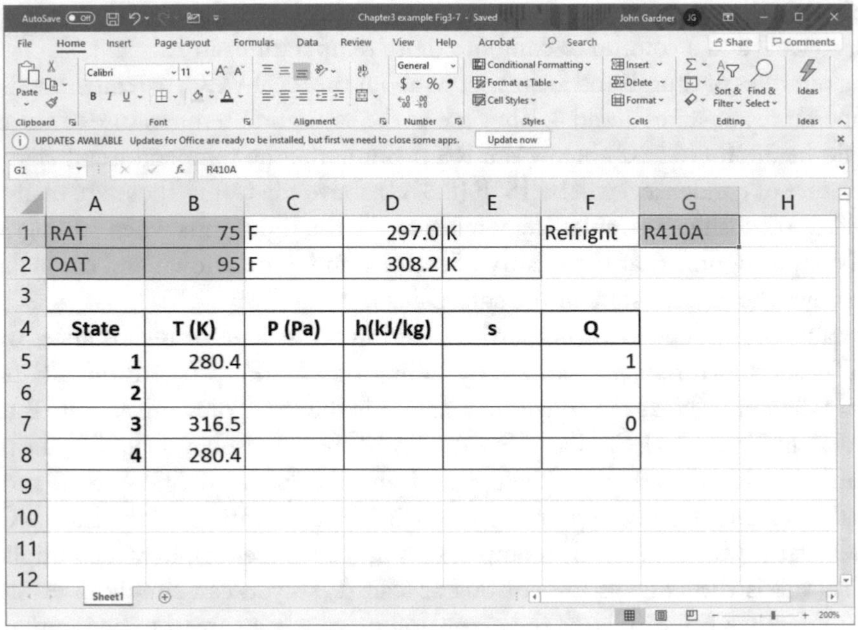

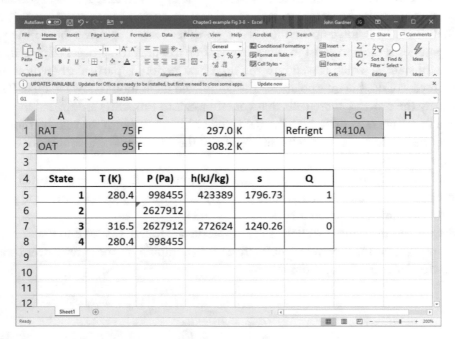

Figure 3.7: Starting point for state properties.

| | A | B | C | D | E | F | G |
|---|---|---|---|---|---|---|---|
| 1 | RAT | 75 | F | 297.0 | K | Refrignt | R410A |
| 2 | OAT | 95 | F | 308.2 | K | | |
| 3 | | | | | | | |
| 4 | State | T (K) | P (Pa) | h(kJ/kg) | s | Q | |
| 5 | 1 | 280.4 | 998455 | 423389 | 1796.73 | 1 | |
| 6 | 2 | | 2627912 | | | | |
| 7 | 3 | 316.5 | 2627912 | 272624 | 1240.26 | 0 | |
| 8 | 4 | 280.4 | 998455 | | | | |

Figure 3.8: Computing the properties of states 1 and 3.

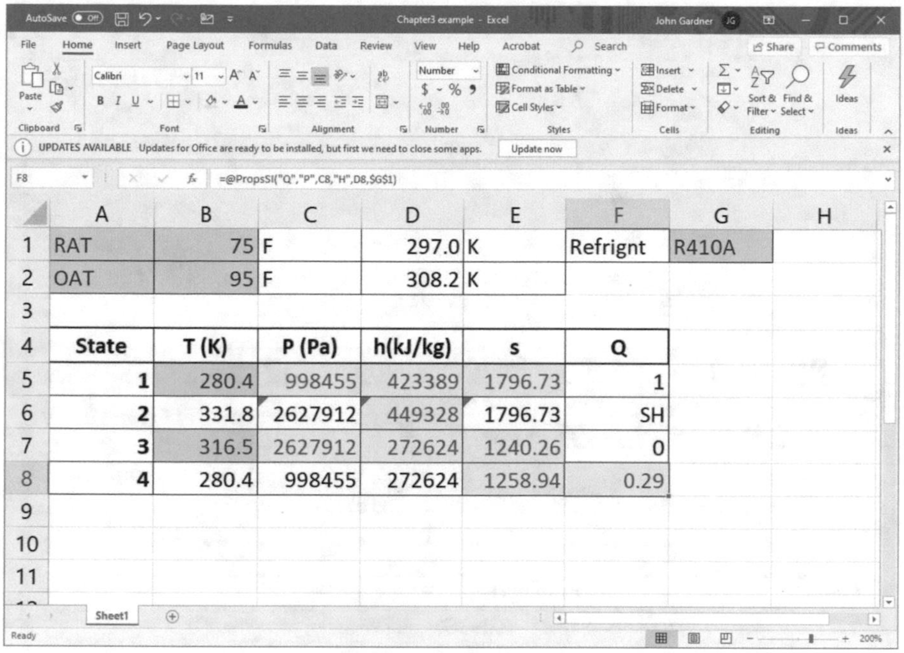


| | A | B | C | D | E | F | G |
|---|---|---|---|---|---|---|---|
| 1 | RAT | 75 | F | 297.0 | K | Refrignt | R410A |
| 2 | OAT | 95 | F | 308.2 | K | | |
| 3 | | | | | | | |
| 4 | State | T (K) | P (Pa) | h(kJ/kg) | s | Q | |
| 5 | 1 | 280.4 | 998455 | 423389 | 1796.73 | 1 | |
| 6 | 2 | 331.8 | 2627912 | 449328 | 1796.73 | SH | |
| 7 | 3 | 316.5 | 2627912 | 272624 | 1240.26 | 0 | |
| 8 | 4 | 280.4 | 998455 | 272624 | 1258.94 | 0.29 | |

Figure 3.9: The properties of all the states are computed.

Table 3.1: AC unit performance specifications; OAT 95F, RAT 75F

| Quantity | Value | Units |
|---|---|---|
| $Q_{low}$ | 21,300 | Btu/hr |
| Power | 1.98 | kW |
| $P_{low}$ | 129 | psig |
| $P_{high}$ | 357 | psig |

Note that the expansion valve is energy-conserving, so these three values must be consistent with the first law, in other words:

$$q_{high} = q_{low} + W_{comp}. \tag{3.8}$$

Now we turn our attention to the unit specifications in Table 3.1. Under these conditions, the unit will remove heat at a rate of 21,300 BTU/h, which is $Q_{low}$. From there it's easy to compute the mass flow rate of the refrigerant as the ratio of the actual heat removed to the heat

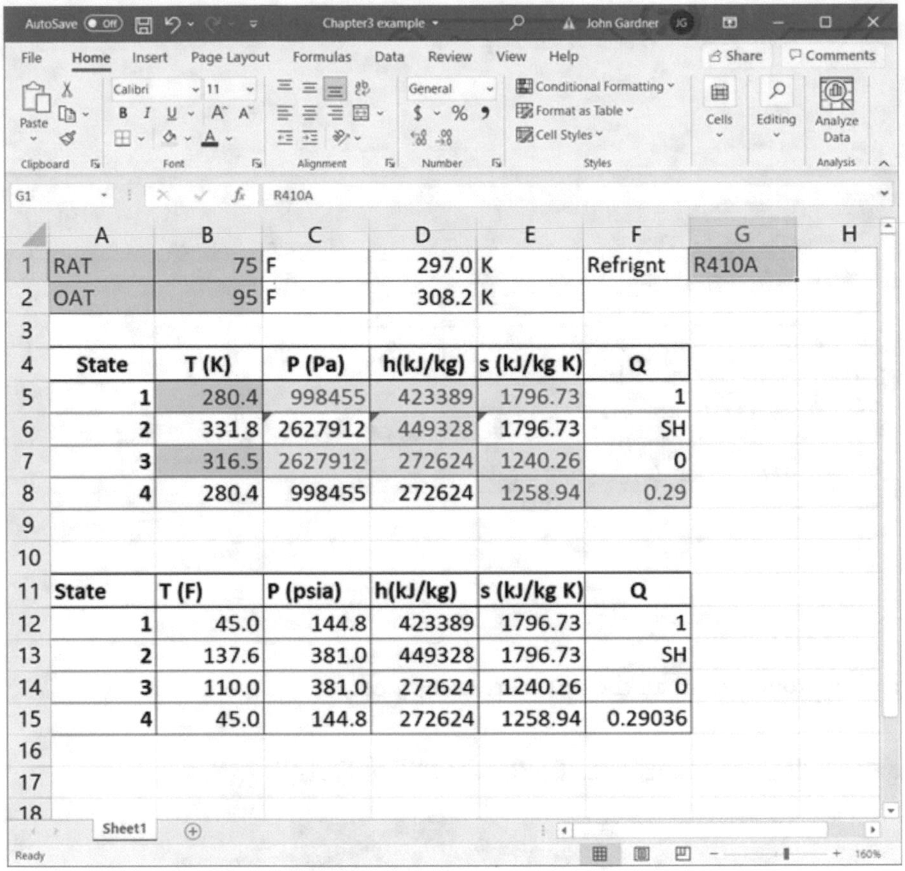

Figure 3.10: Add a state property table with temperature in F and pressure in psi.

removed per unit mass of flow:

$$\dot{m} = \frac{Q_{low}}{q_{low}}. \tag{3.9}$$

Finally, we can compute the energy efficiency ratio (EER) and the coefficient of performance of this cycle. The coefficient of performance is defined in Equation (3.5). The EER is the same ratio, but with the heat transfer expressed in BTU/hr and the energy required to run the unit expressed in W. Figure 3.11 shows the additional tables you can easily add to the spreadsheet to compute these quantities. Finally, we convert the high and low pressure (which are computed in Pa (absolute) to psig to compare to the original specs in the next section.

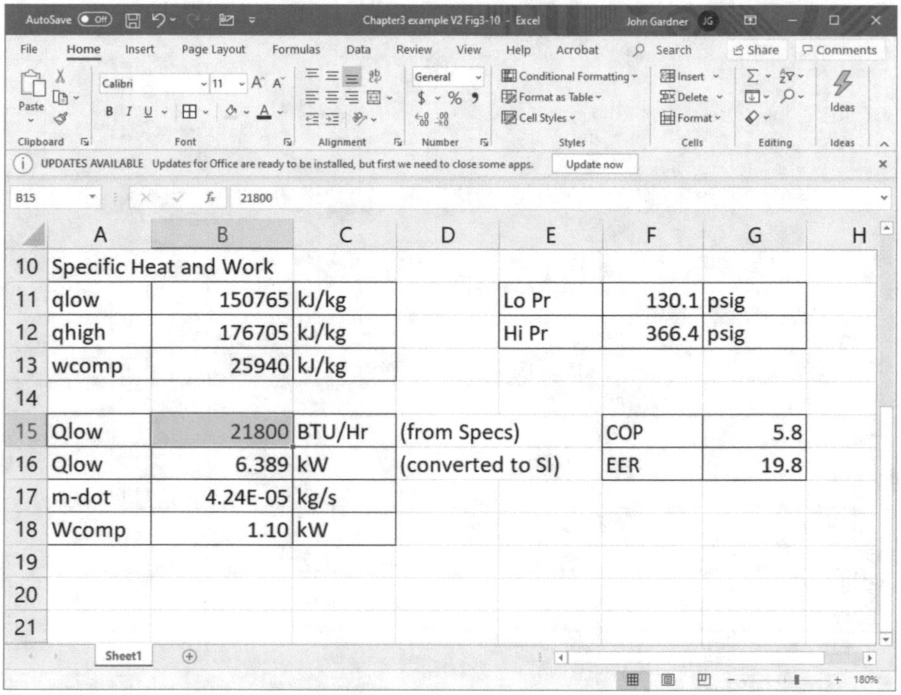

Figure 3.11: The performance characteristics of the cycle.

## 3.4.1    REFINING THE MODEL

Now that the spreadsheet is complete, let's see how well it did. The detailed performance specifications provided by the manufacturer tells us the high and low pressures we could expect under these conditions. Comparing the results in Figure 3.11 with the specifications (Table 3.1), we see that the low pressure we compute is within 1 pisg the specification while the high pressure is about 2.5% higher. Additionally, the COP and EER seem pretty high as we would expect values between 3 and 4 for COP and 11–14 for EER. Where did we go wrong? First, the alignment of the pressures tell us that our assumptions about the saturation temperatures/pressures at the condenser and evaporator were pretty good. Any large discrepancy can be alleviated by adjusting the $\Delta$Ts in cells B5 and B7.

To better understand the COP, it's good to realize that we've also assumed that the compressor was 100% efficient as is the electric motor driving the compressor. We know that neither of these is true. In addition, the compressor efficiency has two components; the isentropic efficiency as well as the volumetric efficiency. The former can be modeled using the proper thermodynamic formula and modifying the spreadsheet, an activity we leave as an exercise for the reader. The volumetric efficiency and the motor efficiency can both be estimated at approximately 90%. Estimating the effect of the isentropic efficiency as approximately 90% as well, we

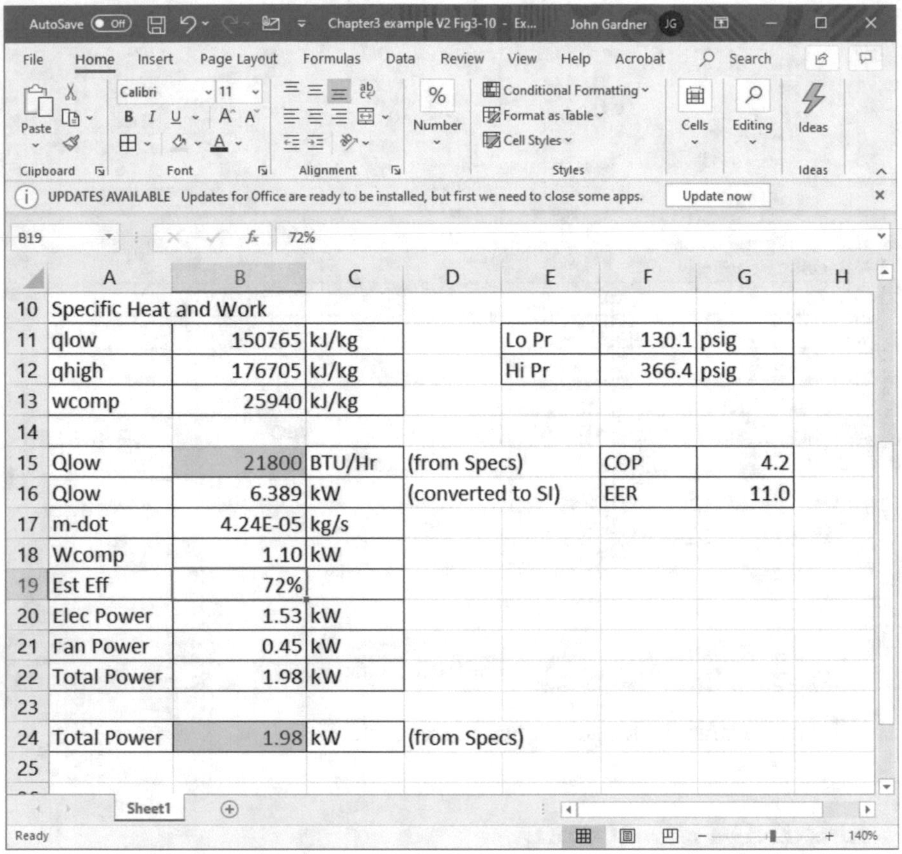

Figure 3.12: Adjusting compressor work to compute realistic performance.

see that the ratio between the ideal compressor work computed in the spreadsheet and the actual electric power delivered to the compressor can computed as the three efficiencies multiplied together, namely 0.9*0.9*0.9 = 73%.

But we're not quite finished yet. The manufacturer's specifications often will include total power draw for given operating conditions. In the case of the example, the unit will draw 1.98 kW. Careful inspection of the supporting documents tells us that this number not only accounts for the compressor work but also the condenser fan (which circulates the outside air to dissipate the heat) and the indoor blower fan. According to the specifications, those motors are rated at 1/8 hp and 1/3 hp, respectively. Converting to SI, the two fans account for approximately 0.45 kW of electric power.

Now we expand the spreadsheet to account for these issue, as shown in Figure 3.12. Cell B19 has our estimate of the overall conversion efficiency from electricity supplied to the compressor to the energy in the refrigerant. B20 is our estimate of the energy supplied to the com-

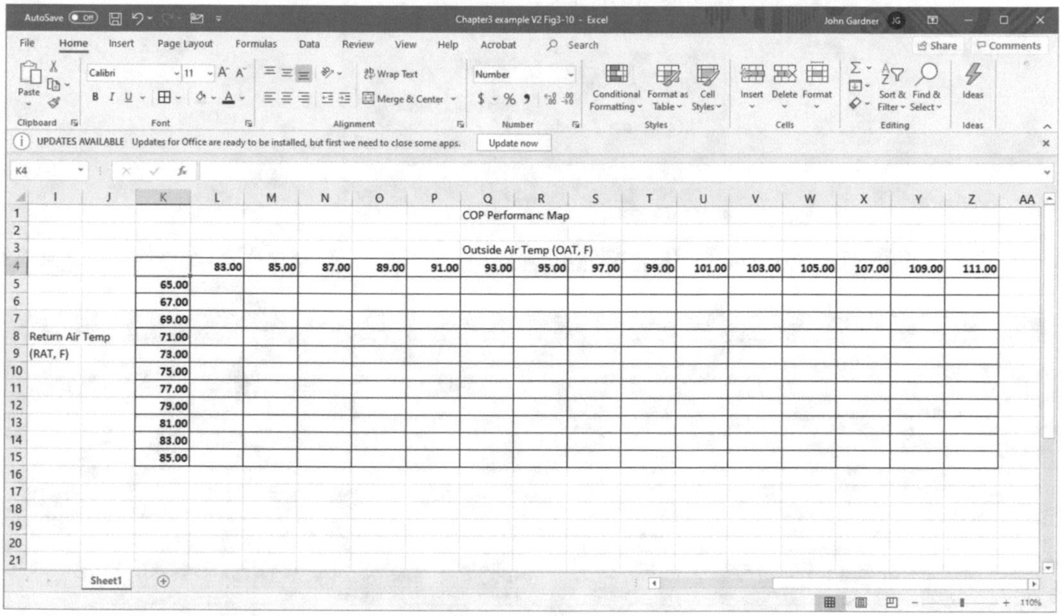

Figure 3.13: Beginning of a COP performance map for 2-ton domestic heat pump using the "What if" functionality in Excel.

pressor motor and B21 is the manufacturer's rating of the power required to run the condenser fan and the blower motors. The thermodynamic analysis estimates total power for the unit to be 1.98 kW which is the same as the number supplied by the manufacturer (B24). Finally, we note that the values for COP and EER are now well inline with expected values for this unit.

## 3.4.2    EXPLORING PERFORMANCE

The expression for the coefficient of performance for the Carnot cycle (Equation (3.6)) tells us that the performance of AC units decreases with increasing OAT and decreasing RAT. Let's use this spreadsheet model to explore that relationship. It would be straightforward yet time-consuming to manually change the OAT and RAT cells one at a time and record the resulting COP in a table. Fortunately, Excel provides an easier way to do this analysis using the "What-If Analysis" tool and data tables. We start by creating the row and column headers representing the temperatures at which we want to evaluate the COP, as shown in Figure 3.13. Next, select the cell in the upper-left corner of the new table and create a formula that references the COP cell (i.e., = G15). Now select the entire table including the headers (K4:Z15) and select the "Data" toolbar at the top of the window and choose "What-If Analysis." In the resulting dialog box, choose "Data Table." Figure 3.14 shows the dialog box. In "Row Input Cell" chose the original cell where we entered the Outside Air Temp (in our case, B2). For "Column Input

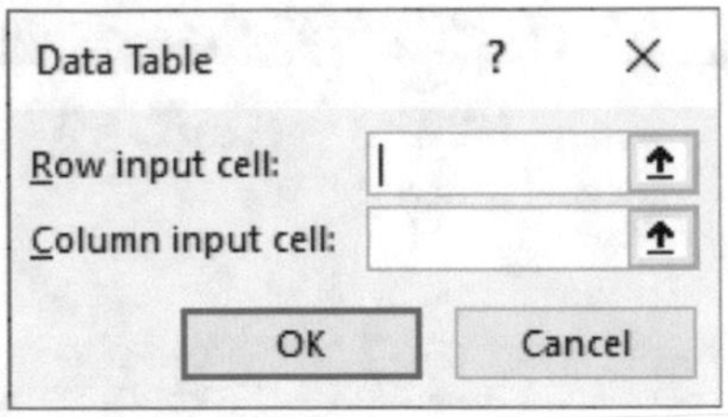


Figure 3.14: Dialog box for the "What if" analysis.

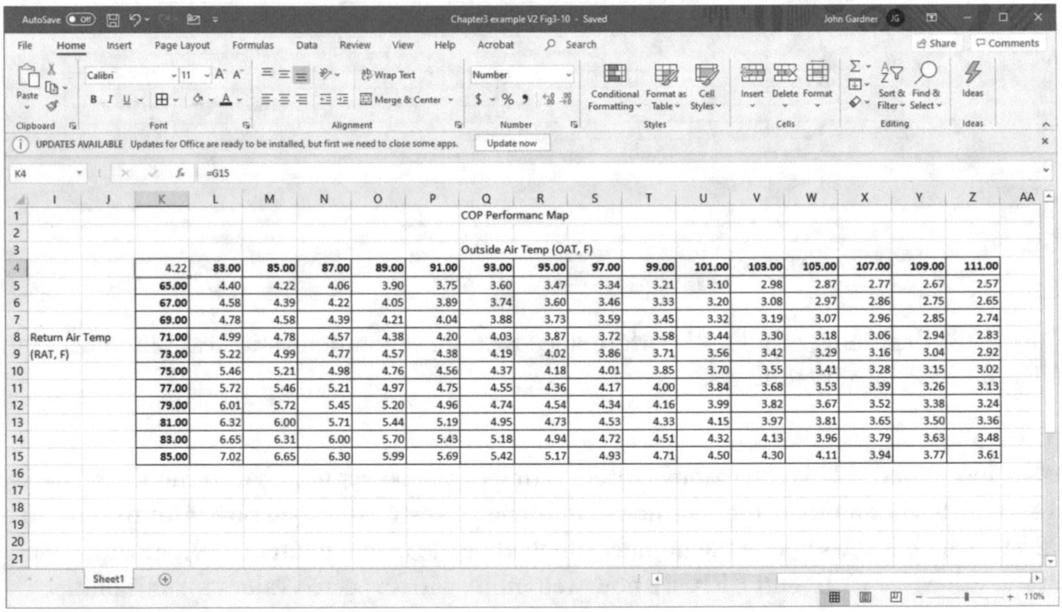


| 4.22 | 83.00 | 85.00 | 87.00 | 89.00 | 91.00 | 93.00 | 95.00 | 97.00 | 99.00 | 101.00 | 103.00 | 105.00 | 107.00 | 109.00 | 111.00 |
|---|---|---|---|---|---|---|---|---|---|---|---|---|---|---|---|
| 65.00 | 4.40 | 4.22 | 4.06 | 3.90 | 3.75 | 3.60 | 3.47 | 3.34 | 3.21 | 3.10 | 2.98 | 2.87 | 2.77 | 2.67 | 2.57 |
| 67.00 | 4.58 | 4.39 | 4.22 | 4.05 | 3.89 | 3.74 | 3.60 | 3.46 | 3.33 | 3.20 | 3.08 | 2.97 | 2.86 | 2.75 | 2.65 |
| 69.00 | 4.78 | 4.58 | 4.39 | 4.21 | 4.04 | 3.88 | 3.73 | 3.59 | 3.45 | 3.32 | 3.19 | 3.07 | 2.96 | 2.85 | 2.74 |
| 71.00 | 4.99 | 4.78 | 4.57 | 4.38 | 4.20 | 4.03 | 3.87 | 3.72 | 3.58 | 3.44 | 3.30 | 3.18 | 3.06 | 2.94 | 2.83 |
| 73.00 | 5.22 | 4.99 | 4.77 | 4.57 | 4.38 | 4.19 | 4.02 | 3.86 | 3.71 | 3.56 | 3.42 | 3.29 | 3.16 | 3.04 | 2.92 |
| 75.00 | 5.46 | 5.21 | 4.98 | 4.76 | 4.56 | 4.37 | 4.18 | 4.01 | 3.85 | 3.70 | 3.55 | 3.41 | 3.28 | 3.15 | 3.02 |
| 77.00 | 5.72 | 5.46 | 5.21 | 4.97 | 4.75 | 4.55 | 4.36 | 4.17 | 4.00 | 3.84 | 3.68 | 3.53 | 3.39 | 3.26 | 3.13 |
| 79.00 | 6.01 | 5.72 | 5.45 | 5.20 | 4.96 | 4.74 | 4.54 | 4.34 | 4.16 | 3.99 | 3.82 | 3.67 | 3.52 | 3.38 | 3.24 |
| 81.00 | 6.32 | 6.00 | 5.71 | 5.44 | 5.19 | 4.95 | 4.73 | 4.53 | 4.33 | 4.15 | 3.97 | 3.81 | 3.65 | 3.50 | 3.36 |
| 83.00 | 6.65 | 6.31 | 6.00 | 5.70 | 5.43 | 5.18 | 4.94 | 4.72 | 4.51 | 4.32 | 4.13 | 3.96 | 3.79 | 3.63 | 3.48 |
| 85.00 | 7.02 | 6.65 | 6.30 | 5.99 | 5.69 | 5.42 | 5.17 | 4.93 | 4.71 | 4.50 | 4.30 | 4.11 | 3.94 | 3.77 | 3.61 |

Figure 3.15: Completed COP performance map for 2-ton domestic heat pump.

Cell," chose the Return Air Temp cell (A2), and hit return. The table should now look like the table in Figure 3.15.

As expected, the COP is highest when the difference between the indoor and outside temperature (i.e., the "lift") is the lowest and becomes quite low when the temperature difference is 46°F. While not ideal, a COP of 2.5 is still considered acceptable for these applications. However, as we'll see in the next chapter, there are applications where the temperature difference can be twice as large and the COP and design challenges of single stage systems for these applications makes this design impractical, thus motivating the consideration of two-stage refrigeration systems.

CHAPTER 4

# Two-Stage Refrigeration Systems

The vapor compression cycle described in Chapter 3 is so useful that it has become ubiquitous in modern society. However, applications in which the difference between the evaporator and condenser temperatures becomes very high requires pressure ratios which are impractical to achieve in compressors at reasonable costs. For these applications, multi-stage compression is often recommended. As you'll recall from other thermodynamic applications, multi-stage cycles present additional opportunities such as intercooling, which greatly enhance overall system efficiency. For the a large portion of industrial refrigeration systems, a two-stage system is implemented. In this chapter, we'll explore the various approaches for two-stage refrigeration systems and discuss the pros and cons of each. In Chapter 5 we'll examine a case study and compare some of these configurations.

## 4.1 CASCADED CYCLES

Figure 4.1 shows the schematic representation of the cascade approach to a two-stage refrigeration system. Note that this entails essentially two separate VCC cycles in which the condenser of the low-pressure cycle is incorporated into a discrete heat exchanger with the evaporator of the high-pressure cycle. The main advantage of this approach is that since the refrigerants of the two cycles never intermingle, different refrigerants can be used in the two systems.

The main disadvantages are that the intermediate heat exchanger can be an expensive addition when compared to other two-stage approaches, and the cascades systems make it more difficult to introduce efficiency measures such as intercoolers.

### 4.1.1 ANALYSIS OF CASCADED REFRIGERATION CYCLES

Analysis of the cascade system is very similar to that of the single-stage system shown in Chapter 3 but the addition of the heat exchanger brings with it some complications, not the least of which being that the mass flow rates in the two systems are likely to be different.

As we typically do when analyzing a thermodynamic cycle, we start with the table of state properties. In this case there are eight states that need to be determined. Next, we fill in the few properties we are likely to know at the beginning of the analysis. Similar to the single stage system in Chapter 3, we start with the temperatures in the evaporator of the low stage and the

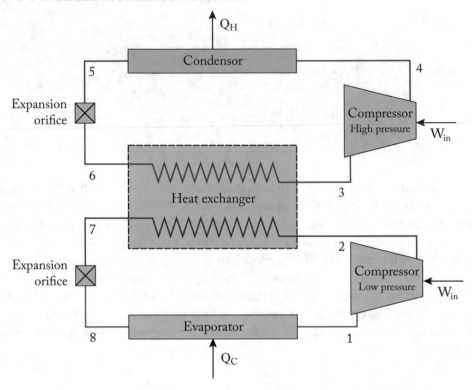

Figure 4.1: Schematic of a cascaded two-stage VCC cycle. The numbers represent the states of the cycle.

condenser of the high stage. The temperatures of states 1 and 8 are going to be 20–30°F lower than the target temperature of the space being cooled, so we'll show those as given. Also, the system will be designed and operated in a way that the temperatures in the heat exchanger are controlled to a desired target. This target temperature can be tuned to maximize the efficiency of the overall system and will be specified as part of the overall design. Therefore, we can assume that temperatures of states 3 and 6 are known. In addition, the temperatures of state 7 will be higher than those of states 3 and 6 by an amount determined by the heat exchanger design and will assumed to be known as well (likely 10–20°F higher.) We will also make the normal assumptions about saturation states in the system and we can start filling out our table as shown in Table 4.1.

We also note that the states entering the compressors (1 and 3) are assumed to be saturated vapor (or a known amount of superheat) and that the states leaving the condensers (5 and 7) are saturated liquid and note those on the table. Since two independent properties completely determine the state, we can now find all properties for states 1, 3, 5, and 7. Hereafter, we'll highlight in blue the entry in the first column for those states that are known.

Table 4.1: State property table for cascaded refrigeration cycles

| State | Temperature | Pressure | Enthalpy | Entropy | Quality |
|---|---|---|---|---|---|
| 1 | $T_{low}$ | | | | 1 |
| 2 | | | | | |
| 3 | $T_{inter}$ | | | | 1 |
| 4 | | | | | |
| 5 | $T_{high}$ | | | | 0 |
| 6 | $T_{inter}$ | | | | |
| 7 | $T_{inter} + \Delta T$ | | | | 0 |
| 8 | $T_{low}$ | | | | |

Table 4.2: Filling out the state property table for cascaded refrigeration cycles

| State | Temperature | Pressure | Enthalpy | Entropy | Quality |
|---|---|---|---|---|---|
| 1 | $T_{low}$ | $P_1$ | $h_1$ | $s_1$ | 1 |
| 2 | | $= P_7$ | | | |
| 3 | $T_{inter}$ | $P_3$ | $h_3$ | $s_3$ | 1 |
| 4 | | $= P_5$ | | | |
| 5 | $T_{high}$ | $P_5$ | $h_5$ | $s_5$ | 0 |
| 6 | $T_{inter}$ | $= P_3$ | $= h_5$ | | |
| 7 | $T_{inter} + \Delta T$ | $P_7$ | $h_7$ | $s_7$ | 0 |
| 8 | $T_{low}$ | $= P_1$ | $= h_7$ | | |

We continue to make the assumption that the pressure drops through the pipes (including the various heat exchangers) are negligible compared to the pressure changes created by the compressors and expansion devices. So the pressures at 1 and 8, 2 and 7, 3, and 6, and 4 and 5 are equal. Finally, note that the expansion orifices are isenthalpic, so we can now fill in the enthalpies for states 6 and 8. These modifications are shown in Table 4.2 and now we can consider states 6 and 8 known.

Next, we observe that we can model the compressors as either isentropic, or having a known isentropic efficiency. In either case, we can now fill in the entropies for states 2 and 4 can be copied from states 1 and 3, respectively, and the table is complete, as shown in Table 4.3.

Table 4.3: Completing the state property table for cascaded refrigeration cycles

| State | Temperature | Pressure | Enthalpy | Entropy | Quality |
|-------|-------------|----------|----------|---------|---------|
| 1 | $T_{low}$ | $P_1$ | $h_1$ | $s_1$ | 1 |
| 2 | | $= P_7$ | | $= s_1$ | |
| 3 | $T_{inter}$ | $P_3$ | $h_3$ | $s_3$ | 1 |
| 4 | | $= P_5$ | | $= s_3$ | |
| 5 | $T_{high}$ | $P_5$ | $h_5$ | $s_5$ | 0 |
| 6 | $T_{inter}$ | $= P_3$ | $= h_5$ | $s_6$ | $Q_6$ |
| 7 | $T_{inter} + \Delta T$ | $P_7$ | $h_7$ | $s_7$ | 0 |
| 8 | $T_{low}$ | $= P_1$ | $= h_7$ | $s_8$ | $Q_8$ |

One last piece of information is needed before the analysis of the system is considered complete. An energy balance of the heat exchanger gives us an important constraint that shows how the mass flow rates of the two cycles are related, as shown in Equation (4.1):

$$\frac{\dot{m}_{low}}{\dot{m}_{high}} = \frac{h_3 - h_6}{h_2 - h_7}. \tag{4.1}$$

As the forgoing discussion indicates, the analysis of a two-stage system can be considerably more involved than that of a single stage and the power and convenience of spreadsheet or other computer tools to aid in the analysis is even more important.

## 4.2   INTERCOOLING WITH A FLASH CHAMBER

The frozen food industry is a dominant user of large-scale refrigeration systems, often keeping large warehouses to −5°F. In these instances, the sheer size of the installations call for refrigerants that are inexpensive and the expense of operating these systems calls for a two-stage approach with intercooling. Most commonly we find a system similar to the schematic shown in Figure 4.2 which makes use of flash chamber and mixing tank. Unlike the cascaded system, there is mixing between the two cycles and hence a common refrigerant must be used. For very large installations, ammonia is the dominant choice. As you can see in this schematic, this is a more complicated arrangement than we've seen before and requires additional discussion.

### 4.2.1   STATES OF A TWO-STAGE SYSTEM WITH FLASH CHAMBER

We'll go through the cycle state by state.

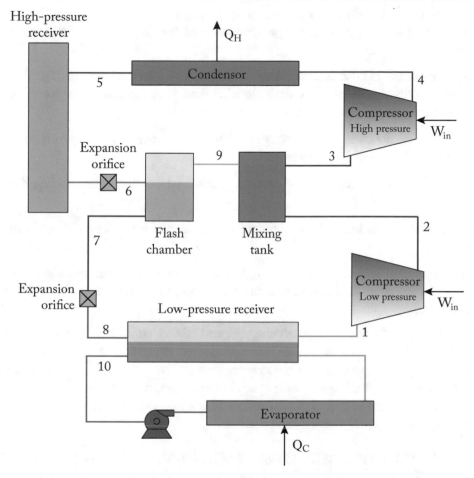

Figure 4.2: Schematic of a two-stage VCC cycle with flash chamber and direct contact heat exchanger (mixing tank). The numbers represent the states of the cycle.

1. Refrigerant enters the low-pressure compressor (also known as the "booster") as saturated (or slightly superheated) vapor at the temperature of the low end of the cycle.

2. The booster compresses to the intermediate pressure and the superheated vapor enters the mixing tank.

3. The mixture (combining streams at states 2 and 9) is superheated vapor and enters the high-pressure compressor.

4. Superheated vapor at the pressure required by the condenser leaves the compressor.

5. The condenser (usually an large rooftop unit using a water spray to aid in heat removal) condenses the refrigerant, leaving saturated liquid (or slightly sub-cooled) to enter the high-pressure receiver.

6. Liquid from the HP Receiver expands through a valve into the flash chamber (which is at the intermediate pressure). The flash chamber contains a saturated mixture of liquid and vapor at the intermediate pressure.

7. The saturated liquid leaves the bottom of the flash chamber and expands to the lower pressure to the low-pressure receiver.

8. Saturated mixture enters the low-pressure receiver where the vapor and liquid are separated.

9. The saturated vapor in the flash chamber flows to the mixing tank providing the intercooling.

10. Saturated liquid at the lower pressure is pumped to the evaporators and allowed to absorb heat from the controlled space. A saturated mixture from the evaporators returns to the LP receiver.

In addition to the use of the flash chamber and mixing tank, there is another significant difference between this configuration and the typical direct expansion approach to refrigeration. In this case, refrigerant expands into the low-pressure receiver and only the saturated liquid is pumped to the various evaporators using special low-temperature ammonia pumps. The saturated mixture leaving the evaporators is introduced back into the LP receiver. Such evaporators are commonly called "liquid overfed" evaporators.

## 4.2.2   THERMODYNAMIC ADVANTAGE OF THE FLASH CHAMBER

To explore the thermodynamic implications of the flash chamber, we turn to the T-S diagram, as seen in Figure 4.3. The states in the T-S diagram refer to the state numbers defined above. In addition, the T-S diagram denoted two additional states, 4' and 6'. These are the hypothetical states that would be found in single-stage VCC system operating between the same to pressures. Now the advantages of the two-stage system with intercooling becomes clearer. First note the area enclosed in the contour defined by 2-3-4-4'-2. That represents the compressor work per unit mass of refrigerant saved when comparing the equivalent single state with the two-stage system. Also note that the state exiting the second expansion valve (state 8) is of considerable lower quality (i.e., more fluid) than the state exiting the single expansion valve in the equivalent system (state 6'). This means that for a given amount of compressor work, there would be considerably less liquid refrigerant available to evaporate in the evaporators, thus decreasing the cooling capacity of the system. Finally, we note that the single-stage system imposes a much larger burden on the condenser units as the gas leaving the single-stage compressor is at a much higher temperature (4') than that of the second-stage compressor (4).

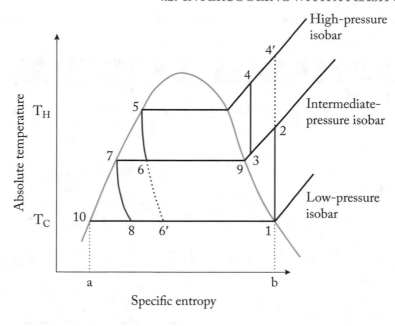

Figure 4.3: T-S diagram of two-stage system with flash changer and mixing tank. The numbers represent the states of the cycle.

## 4.2.3   ANALYSIS OF TWO-STAGE SYSTEM WITH FLASH CHAMBER

We'll follow a procedure similar to that shown in the previous section, by starting with a table of the states and properties around the cycles. In this case, we'll start with the three pressures since all of the states lie on one of the three isobars. The low pressure will be determined by the saturation pressure of the refrigerant at the temperature required in the evaporators (somewhat lower than the temperature in the conditioned space). Similarly, the high pressure is determined by the needs of the condenser. Since these systems normally incorporate evaporative condensers (see Section 1.2.3), the ambient condition we should design to is the wet bulb temperature since that is the temperature approached in the cooling coils of evaporative condensers. The high pressure should be the saturation pressure of the refrigerant at a temperature somewhat higher than the design wet bulb temperature. It should be noted that this pressure can be made to "float" in operating these system, allowing the pressure to achieve the minimum required to adequately drive the heat rejection process in the condensers. This approach can save considerable energy since the alternative is to set the high pressure to accommodate the worst case scenario for the location (i.e., highest outside temperature and highest humidity). The use of floating head pressures to save energy in investigated in the case study presented in Chapter 6.

The intermediate pressure, that of the flash chamber and mixing chamber, is a design variable which may be chosen for to meet various design goals. For example, the liquid at that

Table 4.4: State property table for a two-cycle refrigeration system with intercooler/flash chamber

| State | Temperature | Pressure | Enthalpy | Entropy | Quality |
|-------|-------------|----------|----------|---------|---------|
| 1 | | $P_{low}$ | | | 1 |
| 2 | | $P_{inter}$ | | | |
| 3 | | $P_{inter}$ | | | |
| 4 | | $P_{high}$ | | | |
| 5 | | $P_{high}$ | | | 0 |
| 6 | | $P_{inter}$ | | | |
| 7 | | $P_{inter}$ | | | 0 |
| 8 | | $P_{low}$ | | | |
| 9 | | $P_{inter}$ | | | 1 |
| 10 | | $P_{low}$ | | | 0 |

pressure could be used for a secondary refrigeration system that doesn't require the very low temperatures of the low-pressure loop. Indeed, this is a major advantage of a two-stage system. For example, if the warehouse required spaces at sub-zero temperatures for frozen foods and a higher temperature for non-frozen yet still refrigerated foods, the same system can serve both. But absent the need for a specific temperature at the intermediate level, the pressure can be chosen so that the sum of the work of the two compressor stages is minimized.

States 1, 8, and 10 lie on the low-pressure isobar, states 2, 3, 6, 7, and 9 on the intermediate and 4 and 5 are on the high-pressure isobar. So all pressures are defined. In addition, we can specify the qualities of the states that lie on the saturation dome. States 5, 7, and 10 are saturated liquid (quality 0) while states 1 and 9 are saturated vapor (quality 1). Table 4.4 shows the pressures for all states and quality (and hence all other properties are know for 5 of the 10 states. As in the previous section, once two independent properties are ascertained for a state, that state number will be highlighted in blue.

Now we observe that we have assumed isentropic compression for both stages and hence $s_2 = s_1$ and state 2 is now defined. Similarly, the two expansion valves are isenthalpic so $h_6 = h_5$ and $h_8 = h_7$, which defines states 6 and 8, as is now shown in Table 4.5.

The leaves states 3 and 4 to be determined. For that we will need to analyze the two new elements, the flash chamber and the mixing tank. The next steps require that we differentiate between the mass flow rates in the high-pressure loop and that in the low-pressure loop. The low-

Table 4.5: State property table for a two-cycle refrigeration system with intercooler/flash chamber

| State | Temperature | Pressure | Enthalpy | Entropy | Quality |
|-------|-------------|----------|----------|---------|---------|
| 1 | $T_1$ | $P_{low}$ | $h_1$ | $s_1$ | 1 |
| 2 | | $P_{inter}$ | | $= s_1$ | |
| 3 | | $P_{inter}$ | | | |
| 4 | | $P_{high}$ | | | |
| 5 | $T_5$ | $P_{high}$ | $h_5$ | $s_5$ | 0 |
| 6 | | $P_{inter}$ | $= h_5$ | | |
| 7 | $T_7$ | $P_{inter}$ | $h_7$ | $s_7$ | 0 |
| 8 | | $P_{low}$ | $= h_7$ | | |
| 9 | $T_9$ | $P_{inter}$ | $h_9$ | $s_9$ | 1 |
| 10 | $T_{10}$ | $P_{low}$ | $h_{10}$ | $s_{10}$ | 0 |

pressure mass flow rate will be determined by the refrigeration load as shown in Equation (4.2):

$$\dot{m}_{lp} = \frac{Q_{low}}{(h_1 - h_8)}. \tag{4.2}$$

Now we'll apply the conservation of mass to the flash chamber to see how the two streams split off. We'll designate $\dot{m}_{hp}$ as the mass flow rate in the high-pressure loop. In other words:

$$\dot{m}_{hp} = \dot{m}_3 = \dot{m}_6. \tag{4.3}$$

At the flash chamber, the flow entering at state 6 is a saturated mixture. The liquid part settles to the bottom of the chamber where it is drained off in state 7 while the vapor phase becomes the stream in state 9. To compute the relative magnitude of the two streams, recall the definition of quality as it applies to state 6:

$$x_6 = \frac{m_{6,vapor}}{m_{6,total}}. \tag{4.4}$$

Rearranging and taking the derivative with respect to time yields:

$$\dot{m}_{6,vapor} = x_6 \dot{m}_{6,total} \tag{4.5}$$

which then allows us to make the following conclusions:

$$\dot{m}_9 = x_6 \dot{m}_6 = x_6 \dot{m}_{hp} \tag{4.6}$$

Table 4.6: State property table for a two-cycle refrigeration system with intercooler/flash chamber

| State | Temperature | Pressure | Enthalpy | Entropy | Quality |
|-------|-------------|----------|----------|---------|---------|
| 1 | $T_1$ | $P_{low}$ | $h_1$ | $s_1$ | 1 |
| 2 | $T_2$ | $P_{inter}$ | $h_2$ | $=s_1$ | SH |
| 3 | | $P_{inter}$ | $h_3$ | | |
| 4 | | $P_{high}$ | | $=s_3$ | SH |
| 5 | $T_5$ | $P_{high}$ | $h_5$ | $s_5$ | 0 |
| 6 | $T_6$ | $P_{inter}$ | $= h_5$ | $s_6$ | $x_6$ |
| 7 | $T_7$ | $P_{inter}$ | $h_7$ | $s_7$ | 0 |
| 8 | $T_8$ | $P_{low}$ | $= h_7$ | $s_8$ | $x_8$ |
| 9 | $T_9$ | $P_{inter}$ | $h_9$ | $s_9$ | 1 |
| 10 | $T_{10}$ | $P_{low}$ | $h_{10}$ | $s_{10}$ | 0 |

and

$$\dot{m}_{lp} = \dot{m}_7 = \dot{m}_2 = (1 - x_6)\dot{m}_{hp}. \tag{4.7}$$

Now we can perform an energy balance on the mixing tank:

$$\dot{m}_{hp}h_3 = x_6\dot{m}_{hp}h_9 + (1 - x_6)\dot{m}_{hp}h_2. \tag{4.8}$$

The mass flow rate drops out leaving:

$$h_3 = x_6h_9 + (1 - x_6)h_2 \tag{4.9}$$

which allows us to fill in the remaining properties for state 3.

The final state (4) can be found through the assumption that the second-stage compression is isentropic so $s_4 = s_3$ and our table is complete, as seen in Table 4.6.

## 4.3 TWO-STAGE SYSTEM WITH INTERCOOLER

Another variation on the two stage with intercooling theme is combining the flash chamber and mixing tank into one component, appropriately called an intercooler, as seen in Figure 4.4.

The refrigerant in the intercooler is a saturated liquid/vapor mixture at the intermediate pressure. The superheated vapor exiting the low-pressure compressor (state 2) is introduced to the bottom of the intercooler where the vapor bubbles through the liquid, cooling the vapor and evaporating some of the liquid. In addition, the condensed liquid from the high-pressure

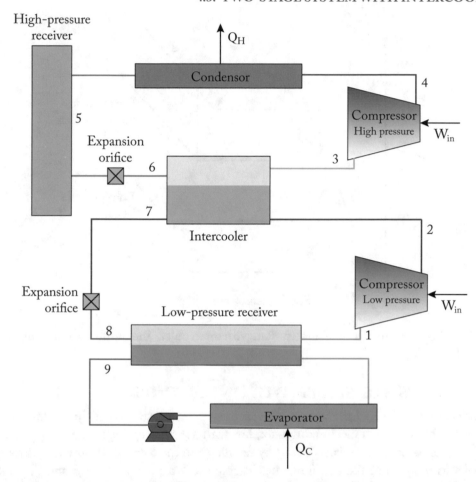

Figure 4.4: Schematic of two-stage system with intercooler. The numbers represent the states of the cycle.

receiver is expanded into the intercooler through an expansion device and enters the intercooler as a saturated mixture (state 6). The saturated liquid is drawn off the bottom of the intercooler to be expanded to the low-pressure receiver while the saturated vapor is drawn off as input to the high-pressure compressor (state 3). As shown in the corresponding T-S diagram, Figure 4.5, this last stream marks the significant difference between this configuration and the previous configuration with the flash chamber/mixing tank. In this case, the input to the high-pressure compressor is saturated, in the former case, it's still superheated. This increases the energy saved for the high-pressure compression. However, it's difficult to make simplistic comparisons as the division of the mass flowrates are different when comparing the two cases, as we will see in the next section.

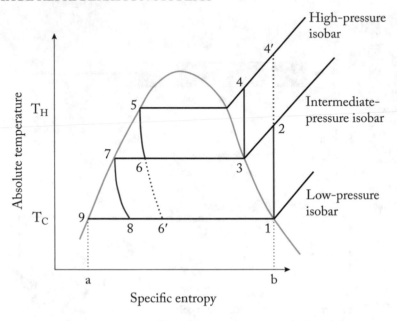

Figure 4.5: T-S diagram of a two-stage system with intercooler. The numbers represent the states of the cycle.

## 4.3.1   ANALYSIS OF SYSTEM WITH INTERCOOLER

Thermodynamic analysis of the system with an intercooler proceeds in a similar fashion as the system with the flash chamber/mixing tank, but with some exceptions. Specifically the intercooler requires its own analysis. We start by defining the the total mass flow rate through the intercooler as the sum of the two flow rates leaving the intercooler (states 3 and 7). Since the two streams are saturated vapor (3) and saturated liquid (7), their ratio is defined by the quality of the mixture in the intercooler.

$$\dot{m} = x\dot{m} + (1 - x)\dot{m}. \tag{4.10}$$

Note that the flow of the vapor out of the intercooler (state 3) is $x\dot{m}$ and the flow of the liquid out of the intercooler (state 7) is $(1 - x)\dot{m}$. Then by continuity (assuming steady state behavior), the other two flows are also $x\dot{m}$ (state 6) and $(1 - x)\dot{m}$ (state 2). Applying conservation of energy to the intercooler gives us:

$$x\dot{m}h_6 + (1 - x)\dot{m}h_2 = x\dot{m}h_3 + (1 - x)\dot{m}h_7. \tag{4.11}$$

Since the intermediate pressure is considered a given in this analysis, we can find the enthalpies at all the states from the analysis as outlined in the previous section. Solving for $x$:

$$x = \frac{h_7 - h_2}{h_6 + h_7 - h_2 - h_3}. \tag{4.12}$$

Summarizing:

$$\dot{m}_{lp} = \frac{Q_{low}}{(h_1 - h_8)} \tag{4.13}$$

and

$$\dot{m}_{hp} = \frac{x}{(1 - x)}\dot{m}_{lp}. \tag{4.14}$$

## 4.4 TWO-STAGE SYSTEM WITH INTERCOOLER AND SUBCOOLER

The final configuration to consider is yet another variation on intercooling. Figure 4.6 shows the schematic in which the flow from the high-pressure receiver is split into two streams, one expanding into the intercooler through an orifice as before (state 6), the other continuing through a heat exchanger that is inside the intercooler, but does not allow mixing of the streams (exiting as state 7). The other two streams are the same as the configuration in the previous section. This arrangement has the effect of sub-cooling the liquid leaving the high-pressure receiver in addition to de-superheating the vapor leaving the low-pressure compressor, both of which have the effect of increasing efficiency. The amount of flow diverted to state 6 into the intercooler is controlled by a level sensor in the intercooler, thus assuring that the refrigerant in the intercooler remains a saturated liquid/vapor mixture. Figure 4.7 shows the resulting T-S diagram.

### 4.4.1 ANALYSIS OF SYSTEM WITH INTERCOOLER AND SUBCOOLER

The analysis of this configuration follows a similar path as the previous sections so we won't got through the details with the exception of the subcooler. As in the previous two configurations, the pressure of the intercooler is a significant design and/or operational variable which may be constrained by other design goals or tuned to achieve maximum efficiency. In addition, the amount of sub-cooling achieved by the subcooler will depend not only on the intercooler pressure (which sets the temperature of the refrigerant in the intercooler) but also on the physical design of the heat exchanger in the intercooler. For the purposes of analysis, we'll assume that the degrees of subcool is known and a characteristic of the subcooler design (and dependent on the intercooler pressure.

The same process outlined in the previous section will allow us to fill out the property table with the exception of state 7. State 7 is determined by subtracting the $\Delta T$ achieved by the subcooler (assumed given) from the temperature of state 5. States 5 and 7 share the same pressure, so the rest of the state 7 properties can now be found. In addition, it is important that we know the portion of the mass flow that is diverted through the expansion orifice to state 6 and into the intercooler, which we'll denote as $x$. Applying an energy balance to the intercooler and designating the mass flow rate through the upper loop as $\dot{m}$, Equation (4.15) gives us:

$$xh_6 + (1 - x)h_2 + (1 - x)(h_5 - h_7) = h_3. \tag{4.15}$$

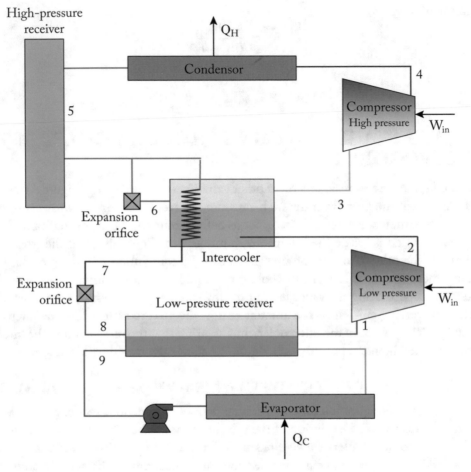

Figure 4.6: T-S diagram of a two-stage system with intercooler/subcooler. The numbers represent the states of the cycle.

Solving for the portion of flow diverted gives us Equation (4.16):

$$x = \frac{h_7 + h_3 - h_5 - h_2}{h_7 + h_6 - h_5 - h_2}. \tag{4.16}$$

Summarizing:

$$\dot{m}_{lp} = \frac{Q_{low}}{(h_1 - h_8)} \tag{4.17}$$

and

$$\dot{m}_{hp} = \frac{1}{(1-x)}\dot{m}_{lp}. \tag{4.18}$$

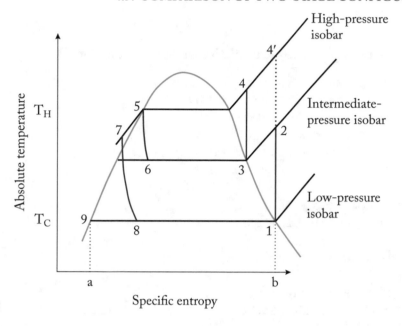

Figure 4.7: T-S diagram of two-stage system with intercooler/subcooler. The numbers represent the states of the cycle.

In the next section, we look at a specific application and compare the performance of the three intercooler configurations.

## 4.5 COMPARISON OF TWO-STAGE CONFIGURATIONS

In this section, we'll preform an analysis of various system configurations operating under the same conditions, as given in Table 4.7.

The configurations we will compare are:

1. Single-Stage VCC

2. Two-Stage VCC with Flash Chamber

3. Two-Stave VCC with Intercooler

4. Two-Stage VCC with Intercooler and Sub-Cooler

To carry out this analysis, we will use a Matlab script for each configuration. For purposes of illustration, the listing for the single-stage configuration is shown here. All four scripts are included in the Appendix. From the analysis, we note the COP for each cycle, the sum of the compressor work for the two compressors, the refrigerant mass flow rates for the low and high-pressure loops, and the theoretical discharge pressure for the high-pressure compressor.

**Listing 4.1** Matlab script to analyze the flash chamber configuration

```
% Script to compare various ammonia cycles
% Set up the baseline parameters for our case
% Case 1: Single Stage, no intercooling
REFR = 'R717'; % Ammonia
ColdTemp = -5.0; % Temperature of cold space in F
OutsideWB = 80.0; % Worst Case WB in F
%
EvapSatTemp = ColdTemp - 30.;
CondensSatTemp = OutsideWB + 15;
%
Qlow = 100.0; % Tons Refrigeration
% convert to SI
EvapSatTempSI = (EvapSatTemp-32)/1.8+273.15; % K
CondensSatTempSI = (CondensSatTemp-32)/1.8+273.15; %K
QlowSI = Qlow*3.517; %kW
% Compute State Properties
P1 = py.CoolProp.CoolProp.PropsSI('P','T',EvapSatTempSI,'Q',1,REFR);
P3 = py.CoolProp.CoolProp.PropsSI('P','T',CondensSatTempSI,'Q',0,REFR);
P2 = P3;
P4 = P1;
%
s1 = py.CoolProp.CoolProp.PropsSI('S','T',EvapSatTempSI,'Q',1,REFR);
s2 = s1; %assume isentropic compression
%
h1 = py.CoolProp.CoolProp.PropsSI('H','T',EvapSatTempSI,'Q',1,REFR);
h2 = py.CoolProp.CoolProp.PropsSI('H','P',P2,'S',s2,REFR);
h3 = py.CoolProp.CoolProp.PropsSI('H','T',CondensSatTempSI,'Q',0,REFR);
h4 = h3;
%
T2 = py.CoolProp.CoolProp.PropsSI('T','P',P2,'S',s2,REFR);
% Compute Energy Values
qlow = h1 - h4;
mdot=QlowSI/qlow;
Wdot = mdot*(h2-h1);
COP = QlowSI/Wdot;
```

Table 4.7: Baseline conditions for case study

| Condition | Imperial Units | SI Units |
|-----------|----------------|----------|
| Cold space temp | −5°F | 252.6 K |
| Evap sat temp | −35°F | 235.9 K |
| Outside WB temp | 80°F | 302.6 K |
| Cond sat temp | 100°F | 310.9 K |
| Refr. load | 100 TonsR | 351.7 kW |

Table 4.8: Results of the comparison case study

| System | COP | $\dot{m}_{lp}, kg/s$ | $\dot{m}_{hp}, kg/s$ | $\dot{W}, kW$ | $T_{high}, K$ |
|--------|-----|------------|------------|---------|----------|
| Single-stage | 2.38 | 3.36E-4 | 3.36E-4 | 147.6 | 444.4 |
| Two-stage, FC | 2.62 | 2.87E-4 | 3.34E-4 | 134.5 | 432.5 |
| Two-stage, IC | 2.77 | 2.86E-4 | 3.75E-4 | 126.9 | 361.6 |
| Two-stage IC+SC | 2.57 | 3.28E-4 | 3.86E-4 | 136.7 | 360.3 |

Table 4.8 summarizes the results computed by the four scripts. The most obvious difference is the COP, which is highest for the intercooler configuration (2.77) as compared to the single stage, which yields a COP of 2.38. This represents a 14% reduction in compressor work for the same amount of refrigeration load. Note that all three of the two-stage configurations result in significant energy reductions. The mass flow rates are indicative of the sizes of the equipment required and the amount of refrigerant that must be used in the system. In this case, the intercooler and intercooler/subcooler configurations appear to require the smallest amount of compressor cooling.

# CHAPTER 5

# Case Study: Comparing Intercoolers

## 5.1    OVERVIEW

In the previous section, we saw four different approaches to two-stage systems, three of which utilized some form of intercooling to achieve better energy efficiency. In this section, we will use the spreadsheet approach to analyze each of the three systems for the same evaporator temperature and refrigeration load. In this analysis we will see how the performance varies as outside conditions and intercooler pressure change. By comparing the coefficient of performance of the systems under various conditions, we can compare their relative strengths and weaknesses.

The conditions for the analysis are described in Table 5.1.

## 5.2    PROCEDURE

To compare and contrast the three types of intercooling described in Chapter 4, we will use the spreadsheet approach first introduced in Chapter 3. The baseline conditions described in Table 5.1 will be used as the starting place and the entire state table computed using the CoolProp add-in for Excel. Once this is accomplished, we will use the "What if Analysis" functionality of Excel to generate a table of the COP for the candidate cycle as a function of outdoor wet bulb temperature and intercooler pressure. This will allow us to assess the optimal intercooler pressure for a given outdoor condition for each of the cycles. Finally, we will compare and contrast the results to examine the pros and cons of each approach.

## 5.3    FLASH CHAMBER

As described in Section 4.2 and depicted schematically in Figure 4.2, this approach to intercooling uses two additional tanks, a flash chamber and a mixing tank. The high-pressure condensed liquid is flashed into the flash tank where the vapor component mixes with the booster compressor outflow in the mixing tank. The liquid component from the flash chamber is further expanded to the low-pressure receiver. The spreadsheet is built up in the same manner as described in Section 4.2 and the resulting screen shot is shown in Figure 5.1. Note that the baseline conditions are shown in the orange-shaded cells. The state table (range A6:I16) shows all the thermodynamic properties for

Table 5.1: Baseline conditions for case study

| Condition | Imperial Units | SI Units |
|---|---|---|
| Cold space temp | –5°F | 252.6 K |
| Evap sat temp | –35°F | 235.9 K |
| Outside WB temp | 80°F | 302.6 K |
| Cond sat temp | 100°F | 310.9 K |
| Refr. load | 100 TonsR | 351.7 kW |

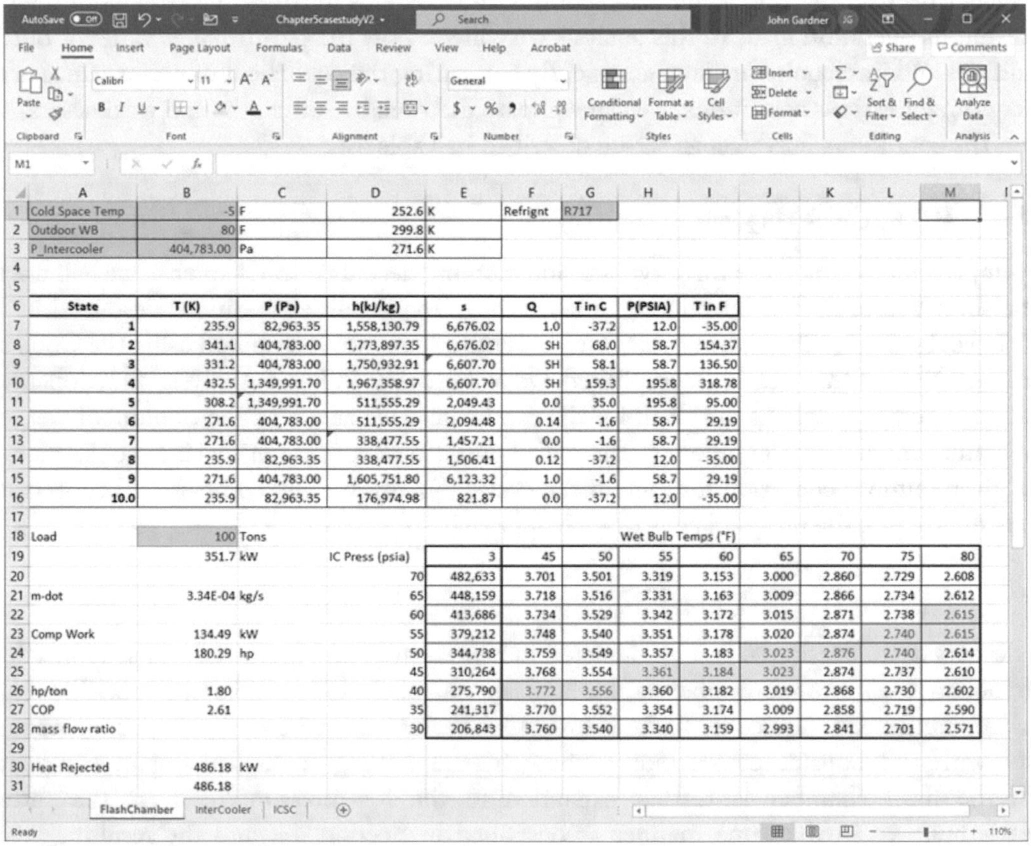

| | | | | | | |
|---|---|---|---|---|---|---|
| Cold Space Temp | -5 | F | 252.6 | K | Refrignt | R717 |
| Outdoor WB | 80 | F | 299.8 | K | | |
| P_Intercooler | 404,783.00 | Pa | 271.6 | K | | |

| State | T (K) | P (Pa) | h(kJ/kg) | s | Q | T in C | P(PSIA) | T in F |
|---|---|---|---|---|---|---|---|---|
| 1 | 235.9 | 82,963.35 | 1,558,130.79 | 6,676.02 | 1.0 | -37.2 | 12.0 | -35.00 |
| 2 | 341.1 | 404,783.00 | 1,773,897.35 | 6,676.02 | SH | 68.0 | 58.7 | 154.37 |
| 3 | 331.2 | 404,783.00 | 1,750,932.91 | 6,607.70 | SH | 58.1 | 58.7 | 136.50 |
| 4 | 432.5 | 1,349,991.70 | 1,967,358.97 | 6,607.70 | SH | 159.3 | 195.8 | 318.78 |
| 5 | 308.2 | 1,349,991.70 | 511,555.29 | 2,049.43 | 0.0 | 35.0 | 195.8 | 95.00 |
| 6 | 271.6 | 404,783.00 | 511,555.29 | 2,094.48 | 0.14 | -1.6 | 58.7 | 29.19 |
| 7 | 271.6 | 404,783.00 | 338,477.55 | 1,457.21 | 0.0 | -1.6 | 58.7 | 29.19 |
| 8 | 235.9 | 82,963.35 | 338,477.55 | 1,506.41 | 0.12 | -37.2 | 12.0 | -35.00 |
| 9 | 271.6 | 404,783.00 | 1,605,751.80 | 6,123.32 | 1.0 | -1.6 | 58.7 | 29.19 |
| 10.0 | 235.9 | 82,963.35 | 176,974.98 | 821.87 | 0.0 | -37.2 | 12.0 | -35.00 |

| | | |
|---|---|---|
| Load | 100 | Tons |
| | 351.7 | kW |
| m-dot | 3.34E-04 | kg/s |
| Comp Work | 134.49 | kW |
| | 180.29 | hp |
| hp/ton | 1.80 | |
| COP | 2.61 | |
| mass flow ratio | | |
| Heat Rejected | 486.18 | kW |
| | 486.18 | |

Wet Bulb Temps (°F)

| IC Press (psia) | 3 | 45 | 50 | 55 | 60 | 65 | 70 | 75 | 80 |
|---|---|---|---|---|---|---|---|---|---|
| 70 | 482,633 | 3.701 | 3.501 | 3.319 | 3.153 | 3.000 | 2.860 | 2.729 | 2.608 |
| 65 | 448,159 | 3.718 | 3.516 | 3.331 | 3.163 | 3.009 | 2.866 | 2.734 | 2.612 |
| 60 | 413,686 | 3.734 | 3.529 | 3.342 | 3.172 | 3.015 | 2.871 | 2.738 | 2.615 |
| 55 | 379,212 | 3.748 | 3.540 | 3.351 | 3.178 | 3.020 | 2.874 | 2.740 | 2.615 |
| 50 | 344,738 | 3.759 | 3.549 | 3.357 | 3.183 | 3.023 | 2.876 | 2.740 | 2.614 |
| 45 | 310,264 | 3.768 | 3.554 | 3.361 | 3.184 | 3.023 | 2.874 | 2.737 | 2.610 |
| 40 | 275,790 | 3.772 | 3.556 | 3.360 | 3.182 | 3.019 | 2.868 | 2.730 | 2.602 |
| 35 | 241,317 | 3.770 | 3.552 | 3.354 | 3.174 | 3.009 | 2.858 | 2.719 | 2.590 |
| 30 | 206,843 | 3.760 | 3.540 | 3.340 | 3.159 | 2.993 | 2.841 | 2.701 | 2.571 |



Figure 5.1: Spreadsheet analysis of the flash chamber/mixing tank configuration.

the ammonia in the cycle using the intercooler pressure indicated in cell B3. The mass flow rate and energy/power computations are listed in the cell range A18:B31.

Using the "What if Analysis" feature in Excel, the table shown in the range E18:M28 is constructed. It shows the COP for the cycle for a range of wet bulb temperatures and intercooler pressures. Note that the optimal intercooler pressure ranges for each wet bulb temperature are indicated with green shading of the highest COP in each column.

From this we make some broad observations. First, the COP ranges from a high of 3.772 at 45°F WB to a low of 2.615 at 80°F WB. Not surprisingly, the lower the outside temperature, the higher the COP. Second, we note that as the WB temperature increases (which necessitates a higher condenser saturation temperature), the optimal intercooler pressure also rises. This is consistent with the theoretical result for multi-stage compression that energy of compression is minimized with the two stages of compression having equal compression ratios. As the outlet of the high-pressure compressor increases, the intercooler pressure that gives rise to that condition must also increase. Finally, we note that the COP becomes less sensitive to changes in intercooler pressure as the wet bulb temperature increases, as indicated by the fact that there is only on pressure at 45°F WB that results in the highest COP (to three decimal places) while there are four such pressures at 75 and 80°F WB.

## 5.4    INTERCOOLER

This section describes the classical intercooler configuration as shown in Figure 4.4. In this case, a single tank is used at the intermediate temperature. The high-pressure liquid is allowed to expand into the tank and the liquid phase passes to the main expansion device to the low-pressure evaporator. In addition, the superheated vapor leaving the booster compressor is introduced at the bottom of the intercooler and allowed to bubble through the liquid phase, thus increasing the heat transfer. The vapor phase in the intercooler feeds the input to the high-pressure compressor.

As described previously, the spreadsheet is built up to define all thermodynamic states at a given condition and the "What if Analysis" analysis is used to build a table of COP as a function of outside WB and intercooler pressure. The results are shown in Figure 5.2.

Our broad observations for the COP table tell us that optimal intercooler pressure is generally much higher for this configuration as compared to the flash chamber in the previous sections. The COP for the intercooler is maximized at intercooler pressures that range from 70–100 psia as opposed to 40–60 psia for the flash chamber. In addition, the COP is about 5% higher for the intercooler configuration, ranging from 3.97 at WB of 44°F to 2.772 at 80°F. And as we noted in the flash chamber case, the COP is less sensitive to intercooler pressure at higher wet bulb temperatures.

**Spreadsheet — InterCooler sheet**

| State | T (K) | P (Pa) | h(kJ/kg) | s | Q | T in C | P(PSIA) | T in F |
|---|---|---|---|---|---|---|---|---|
| 1 | 235.9 | 82,963.35 | 1,558,130.79 | 6,676.02 | 1.0 | -37.2 | 12.0 | -35.00 |
| 2 | 341.1 | 404,783.00 | 1,773,897.35 | 6,676.02 | SH | 68.0 | 58.7 | 154.37 |
| 3 | 271.6 | 404,783.00 | 1,605,751.80 | 6,123.32 | 1.0 | -1.6 | 58.7 | 29.19 |
| 4 | 357.3 | 1,349,991.70 | 1,776,818.85 | 6,123.32 | SH | 84.1 | 195.8 | 183.42 |
| 5 | 308.2 | 1,349,991.70 | 511,555.29 | 2,049.43 | 0.0 | 35.0 | 195.8 | 95.00 |
| 6 | 271.6 | 404,783.00 | 511,555.29 | 2,094.48 | 0.14 | -1.6 | 58.7 | 29.19 |
| 7 | 271.6 | 404,783.00 | 338,477.55 | 1,457.21 | 0.0 | -1.6 | 58.7 | 29.19 |
| 8 | 235.9 | 82,963.35 | 338,477.55 | 1,506.41 | 0.12 | -37.2 | 12.0 | -35.00 |
| 9 | 235.9 | 82,963.35 | 176,974.98 | 821.87 | 0.0 | -37.2 | 12.0 | -35.00 |

Top section:
- Cold Space Temp: -5 F — 252.6 K — Refrignt: R717
- Outdoor WB: 80 F — 299.8 K
- P_Intercooler: 404,783.00 Pa — 271.6 K

What-if analysis section:
- Load: 100 Tons; 351.7 kW
- m-dot,low: 2.88E-04 kg/s
- m-dot,high: 3.78E-04 kg/s
- Comp Work: 126.93 kW; 170.14 hp
- hp/ton: 1.70
- COP: 2.77
- mass flow ratio
- Heat Rejected: 478.61 kW; 478.61

| IC Press (psia) | | 3 | 45 | 50 | 55 | 60 | 65 | 70 | 75 | 80 |
|---|---|---|---|---|---|---|---|---|---|---|
| 110 | | 428,424 | 3.878 | 3.681 | 3.499 | 3.331 | 3.175 | 3.030 | 2.895 | 2.768 |
| 105 | | 408,950 | 3.895 | 3.694 | 3.510 | 3.340 | 3.182 | 3.035 | 2.898 | 2.770 |
| 100 | | 389,476 | 3.911 | 3.707 | 3.520 | 3.347 | 3.188 | 3.039 | 2.901 | 2.772 |
| 95 | | 370,002 | 3.925 | 3.719 | 3.529 | 3.354 | 3.192 | 3.042 | 2.902 | 2.772 |
| 90 | | 350,528 | 3.939 | 3.729 | 3.536 | 3.359 | 3.195 | 3.043 | 2.902 | 2.771 |
| 85 | | 331,055 | 3.950 | 3.737 | 3.542 | 3.362 | 3.197 | 3.043 | 2.901 | 2.768 |
| 80 | | 311,581 | 3.960 | 3.743 | 3.546 | 3.364 | 3.196 | 3.041 | 2.897 | 2.764 |
| 75 | | 292,107 | 3.966 | 3.747 | 3.547 | 3.363 | 3.194 | 3.037 | 2.892 | 2.757 |
| 70 | | 272,633 | 3.970 | 3.748 | 3.545 | 3.359 | 3.188 | 3.031 | 2.885 | 2.749 |
| 65 | | 253,159 | 3.969 | 3.745 | 3.540 | 3.352 | 3.180 | 3.021 | 2.874 | 2.738 |

(Wet Bulb Temps (°F) span the columns 45–80. Sheet tabs: FlashChamber, InterCooler, ICSC.)

Figure 5.2: Spreadsheet analysis of the intercooler configuration.

## 5.5   INTERCOOLER WITH SUBCOOLER

The last configuration to consider is the intercooler with an integrated subcooler as shown in Figure 4.4. This is the most complicated of the approaches and has an additional design parameter to consider, namely the amount of subcool that that internal heat exchanger is designed to achieve. For this exercise, we assume the subcooler achieves a $\Delta T$ of 10°F. Figure 5.3 shows the spreadsheet with the state properties and the "What if Analysis." Note that the trends we see in the COP table are similar to those in for the flash chamber.

The addition of the subcooler makes this particular configuration a bit more flexible for applications which require an intermediate pressure evaporator system, which is common in applications requiring both refrigeration and freezing. In these cases, the intermediate pressure is determined by the desired temperature of the refrigeration system, often requiring a sub-optimal choice of intercooler pressure. However, the subcooler in the intercooler can be designed

| Cold Space Temp | -5 | F | 252.6 | K | Refrignt | R717 |
| Outdoor WB | 80 | F | 299.8 | K | | |
| P_Intercooler | 460,000.00 | Pa | 275.0 | K | | |
| DT Subcooler | 10 | F | 5.6 | K | | |

| State | T (K) | P (Pa) | h(kJ/kg) | s | Q | T in C | P(PSIA) | T in F |
|---|---|---|---|---|---|---|---|---|
| 1 | 235.9 | 82,963.35 | 1,558,130.79 | 6,676.02 | 1.0 | -37.2 | 12.0 | -35.00 |
| 2 | 351.1 | 460,000.00 | 1,794,951.92 | 6,676.02 | SH | 78.0 | 66.7 | 172.37 |
| 3 | 275.0 | 460,000.00 | 1,609,427.31 | 6,078.64 | 1.0 | 1.9 | 66.7 | 35.36 |
| 4 | 351.3 | 1,349,991.70 | 1,760,989.33 | 6,078.64 | SH | 78.2 | 195.8 | 172.69 |
| 5 | 308.2 | 1,349,991.70 | 511,555.293 | 2,049.43 | 0.0 | 35.0 | 195.8 | 95.00 |
| 6 | 275.0 | 460,000.00 | 511,555.293 | 2,086.59 | 0.13 | 1.9 | 66.7 | 35.36 |
| 7 | 302.6 | 1,349,991.70 | 484,631.369 | 1,961.26 | 0.0 | 29.4 | 195.8 | 85.00 |
| 8 | 235.9 | 82,963.35 | 484,631.369 | 2,125.90 | 0.22 | -37.2 | 12.0 | -35.00 |
| 9 | 235.9 | 82,963.35 | 176,974.98 | 821.87 | 0.0 | -37.2 | 12.0 | -35.00 |

| Load | 100 | Tons | | Wet Bulb Temps (°F) | | | | | | | | |
|---|---|---|---|---|---|---|---|---|---|---|---|---|
| | 351.7 | kW | IC Press (psia) | 3 | 45 | 50 | 55 | 60 | 65 | 70 | 75 | 80 |
| x | 0.16 | | 70 | 482,633 | 3.742 | 3.530 | 3.335 | 3.157 | 2.992 | 2.840 | 2.699 | 2.568 |
| m-dot,low | 3.28E-04 | kg/s | 65 | 448,159 | 3.755 | 3.540 | 3.344 | 3.163 | 2.998 | 2.844 | 2.703 | 2.571 |
| m-dot,high | 3.91E-04 | kg/s | 60 | 413,686 | 3.767 | 3.550 | 3.351 | 3.169 | 3.002 | 2.848 | 2.705 | 2.572 |
| Comp Work | 136.84 | kW | 55 | 379,212 | 3.778 | 3.558 | 3.357 | 3.174 | 3.005 | 2.850 | 2.706 | 2.572 |
| | 183.44 | hp | 50 | 344,738 | 3.787 | 3.564 | 3.362 | 3.177 | 3.007 | 2.850 | 2.706 | 2.571 |
| | | | 45 | 310,264 | 3.793 | 3.568 | 3.364 | 3.177 | 3.006 | 2.849 | 2.703 | 2.568 |
| hp/ton | 1.83 | | 40 | 275,790 | 3.795 | 3.568 | 3.362 | 3.174 | 3.002 | 2.844 | 2.698 | 2.563 |
| COP | 2.57 | | 35 | 241,317 | 3.791 | 3.563 | 3.356 | 3.167 | 2.994 | 2.835 | 2.689 | 2.554 |
| | | | 30 | 206,843 | 3.779 | 3.550 | 3.342 | 3.153 | 2.980 | 2.821 | 2.675 | 2.539 |
| Heat Rejected | 488.53 | kW | | | | | | | | | | |
| | 488.53 | | | | | | | | | | | |

Sheet tabs: FlashChamber | InterCooler | ICSC

Figure 5.3: Spreadsheet analysis of the configuration with subcooler integrated in the intercooler.

to increase efficiency. In general, the spreadsheet analysis shows that increasing subcool (i.e., increasing the heat transfer area of the embedded heat exchanger) will increase COP.

## 5.6 SUMMARY

In this chapter, we explored the use of the integration of CoolProp with advanced spreadsheet methods to illustrate how these tools could be used for in-depth analysis of complex thermodynamic cycles. In addition to the "What if Analysis" data table tool, other advanced methods include goal seeking and optimization algorithms that a design engineer would find invaluable. In the next chapter, we will use similar tools to analyze cycle behavior throughout the year using weather databases specific to your location.

# CHAPTER 6

# Case Study: Discharge Pressure

## 6.1   OVERVIEW

One of the take-home messages of the previous chapters is that higher outside temperatures lower overall efficiencies. The underlying mechanism is clear because as outside temperatures increase, the outlet compressor pressure must also increase to provide adequate $\Delta T$ to drive the heat exchange from the refrigerant to the environment. That $\Delta T$ (the refrigerant saturation temperature minus the outside wet bulb temperature) is known as the *approach* for the condenser. Industrial evaporative condensers are designed for an approach of 8–12°F (4.5–6.5°C). Therefore, some systems are set to operate at a compressor discharge pressure that is adequate to maintain the design approach for the worst case scenario, i.e., the worst condition likely to be encountered at that location. The result is a system that operates the majority of the time at a pressure that's much higher than required and hence less efficient than is possible.

As we will discuss subsequently, there may be other reasons to set the discharge pressure, but achieving the proper temperature in the condenser is the dominant reason. Whatever the reasons for high-pressure setting on the discharge, it is instructive to look at the impact of reducing the pressure to only what is needed for a given situation. In this case study, we'll examine the use of the Typical Meteorological Year (TMY) data sets to estimate cost savings from allowing the discharge pressure to "float." That is, the pressure will adapt to the condenser conditions to maintain the proper approach. This is one of the most common recommendations to improve the efficiency of industrial refrigeration systems [WMB04].

## 6.2   BASELINE CASE: FIXED DISCHARGE PRESSURE

As a baseline, we'll use a case similar to our standard conditions in Chapter 5. We'll assume 100 tons of refrigeration load, a cold space temperature of −5°F, and a fixed discharge pressure of 175 psia. The discharge pressure sets the condenser saturation temperature at 88°F. This would be a good approximation of a worst-case scenario in most of the United States.

Using the same methodology outlined in Chapter 4 and using the intercooler configuration shown in Figure 4.4, we find that the system has a theoretical compressor work of 118.9 kW when the intercooler pressure is set to 51 psia (350 kPa). Since frozen food warehouses typically operate around the clock and year round, we can estimate the amount of electricity it would take to provide that compressor work. Using the rough estimate of 70% electrical-to-compressor efficiency, we can estimate that roughly 1,490,000 kWh per year of operation. At a rate of $0.066

per kilowatt hour (which was the average cost for industrial U.S. customers in 2020[1]), the cost would be roughly $98,200 per year for every 100 tons of refrigeration capacity. This will be the baseline against which we will compare the results of the following sections.

## 6.3 TMY FILES, WEB BULB TEMPS, AND DATA BINS

In order to support a wide range of engineering activities from solar photovoltaic system design to building energy modeling, the National Renewable Energy Lab maintains an extensive database of typical weather conditions for thousands of locations around the world. A common method of accessing this database is through Typical Meteorological Year (TMY) files.[2] TMY files are spreadsheet-compatible which contain a row for each hour in the year. Each month of those data are real data from a year which, according to the statistical analysis, represents the most typical month of the 30 years on record. Therefore, the data represent actual measurements at the location, but the January data may come from one year, the February from another, and so on. There is also some smoothing that avoids large discontinuities at the transitions between the months. Among other measurements, TMY files have the dry bulb temperature, dew point, pressure, and solar insolation measurements.

As noted in previous chapters, the evaporative condensers often used in industrial applications increase heat transfer by lowering the local temperature outside the coils to a point near the web bulb temperature, which can be as much as 30°F lower than the dry bulb temperature. Therefore, it would be very useful to find the wet bulb temperature from the TMY data. As noted in Chapter 2, CoolProp has a function specifically targeted to finding properties of humid air.

Referring to Table 2.2 for the property labels, we see that the wet bulb temperature can be computed from the dry bulb temperature, atmospheric pressure, and dew point. The following example shows this functionality for a very hot and dry day where the dry bulb temp is 102°F (39°C), the dew point is 5°C and the pressure is 91,000 Pa.

```
=HAPropsSI("Twb","T",39+273.15,""P",91000,"D",5+273.15)-273.15.
```

The value returned in this case is 18.183°C (64.7°F).

Starting with a TMY file, we will need to add a column to the TMY files and use the CoolProp plug-in to compute the wet bulb temperature that corresponds to the temperature, dew point and pressure for each hour in the data set. Figure 6.1 shows a screen shot of the TMY file for Atlanta, Georgia. Note that column K in the spreadsheet is not part of the original TMY file. It was inserted and the HAPropsSI function was used to compute the wet bulb temperature

---

[1]https://www.statista.com/statistics/190680/us-industrial-consumer-price-estimates-for-retail-electricity-since-1970/
[2]https://nsrdb.nrel.gov/about/tmy.html

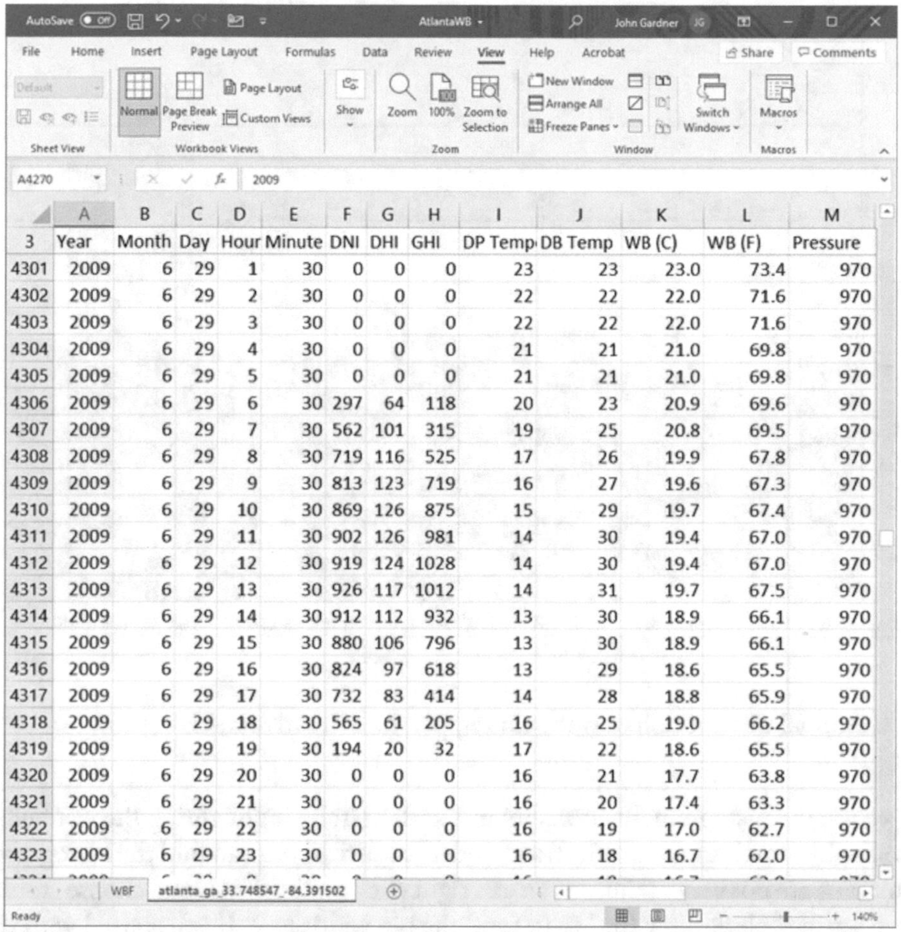

| 3 | Year | Month | Day | Hour | Minute | DNI | DHI | GHI | DP Temp | DB Temp | WB (C) | WB (F) | Pressure |
|---|---|---|---|---|---|---|---|---|---|---|---|---|---|
| 4301 | 2009 | 6 | 29 | 1 | 30 | 0 | 0 | 0 | 23 | 23 | 23.0 | 73.4 | 970 |
| 4302 | 2009 | 6 | 29 | 2 | 30 | 0 | 0 | 0 | 22 | 22 | 22.0 | 71.6 | 970 |
| 4303 | 2009 | 6 | 29 | 3 | 30 | 0 | 0 | 0 | 22 | 22 | 22.0 | 71.6 | 970 |
| 4304 | 2009 | 6 | 29 | 4 | 30 | 0 | 0 | 0 | 21 | 21 | 21.0 | 69.8 | 970 |
| 4305 | 2009 | 6 | 29 | 5 | 30 | 0 | 0 | 0 | 21 | 21 | 21.0 | 69.8 | 970 |
| 4306 | 2009 | 6 | 29 | 6 | 30 | 297 | 64 | 118 | 20 | 23 | 20.9 | 69.6 | 970 |
| 4307 | 2009 | 6 | 29 | 7 | 30 | 562 | 101 | 315 | 19 | 25 | 20.8 | 69.5 | 970 |
| 4308 | 2009 | 6 | 29 | 8 | 30 | 719 | 116 | 525 | 17 | 26 | 19.9 | 67.8 | 970 |
| 4309 | 2009 | 6 | 29 | 9 | 30 | 813 | 123 | 719 | 16 | 27 | 19.6 | 67.3 | 970 |
| 4310 | 2009 | 6 | 29 | 10 | 30 | 869 | 126 | 875 | 15 | 29 | 19.7 | 67.4 | 970 |
| 4311 | 2009 | 6 | 29 | 11 | 30 | 902 | 126 | 981 | 14 | 30 | 19.4 | 67.0 | 970 |
| 4312 | 2009 | 6 | 29 | 12 | 30 | 919 | 124 | 1028 | 14 | 30 | 19.4 | 67.0 | 970 |
| 4313 | 2009 | 6 | 29 | 13 | 30 | 926 | 117 | 1012 | 14 | 31 | 19.7 | 67.5 | 970 |
| 4314 | 2009 | 6 | 29 | 14 | 30 | 912 | 112 | 932 | 13 | 30 | 18.9 | 66.1 | 970 |
| 4315 | 2009 | 6 | 29 | 15 | 30 | 880 | 106 | 796 | 13 | 30 | 18.9 | 66.1 | 970 |
| 4316 | 2009 | 6 | 29 | 16 | 30 | 824 | 97 | 618 | 13 | 29 | 18.6 | 65.5 | 970 |
| 4317 | 2009 | 6 | 29 | 17 | 30 | 732 | 83 | 414 | 14 | 28 | 18.8 | 65.9 | 970 |
| 4318 | 2009 | 6 | 29 | 18 | 30 | 565 | 61 | 205 | 16 | 25 | 19.0 | 66.2 | 970 |
| 4319 | 2009 | 6 | 29 | 19 | 30 | 194 | 20 | 32 | 17 | 22 | 18.6 | 65.5 | 970 |
| 4320 | 2009 | 6 | 29 | 20 | 30 | 0 | 0 | 0 | 16 | 21 | 17.7 | 63.8 | 970 |
| 4321 | 2009 | 6 | 29 | 21 | 30 | 0 | 0 | 0 | 16 | 20 | 17.4 | 63.3 | 970 |
| 4322 | 2009 | 6 | 29 | 22 | 30 | 0 | 0 | 0 | 16 | 19 | 17.0 | 62.7 | 970 |
| 4323 | 2009 | 6 | 29 | 23 | 30 | 0 | 0 | 0 | 16 | 18 | 16.7 | 62.0 | 970 |

Figure 6.1: Spreadsheet screenshot of TMY file with wet bulb columns inserted.

from the temperature, pressure, and dew point data given for that row. We also add column L to show the wet bulb temperature in °F.

In this figure, we used the "freeze pane" function in Excel to keep the headers in place and scrolled down to the very warm days in July to see just how much the wet bulb temperature can be below the dry bulb. This TMY file shows a temperature of 30°C (86°F) and a wet bulb temperature of 18.9°C (66°F). In the dryer climates of the western U.S., the wet bulb temperature depression can be even greater.

To use these data in our analysis, the wet bulb temperature plus the approach sets the saturation temperature of the condenser, which in turn sets the discharge pressure for the high-stage compressor. To carry out the analysis, as described in the next section, we first need to "bin" the data to find out how many hours a year the wet bulb temperature lies in various ranges.

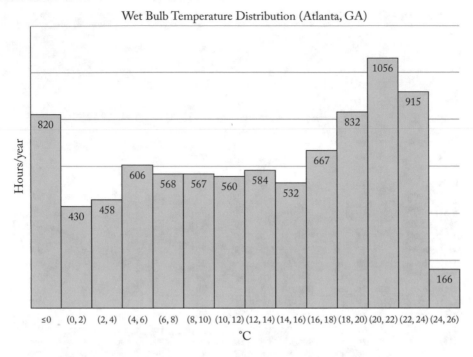

Figure 6.2: Annual distribution of wet bulb temperature in Atlanta, GA (°C).

For this, we use the histogram functionality in Excel. Start by highlighting the wet bulb column then go to the insert>chart menu selection. Under "statistics" you can find the histogram option. The default bins are not very useful, so you can right click on the $x$-axis and choose "format axis." From there set the "Underflow Bin" to 0.0 and the bin width to 2. If you started with the TMY file for Atlanta, you should have a chart that looks like the chart in Figure 6.2.

Since the refrigeration industry practitioners often prefer to work in Imperial units, this histogram is easily modified to present the same data in Fahrenheit, as shown in Figure 6.3.

## 6.4   ESTIMATING COST SAVINGS WITH LOWER DISCHARGE PRESSURE

We have established that if the compressor discharge pressure is set to accommodate the highest wet bulb temperature in a given location, it is wasting energy for much of the time because as wet bulb temperature drops, the condenser saturation temperature, and hence the discharge pressure can also fall. However, there are some factors that must be considered before we get too aggressive in reducing discharge pressure.

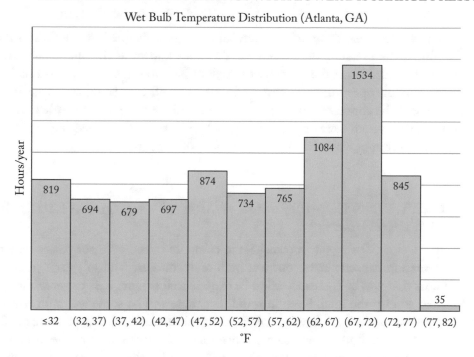

Figure 6.3: Annual distribution of wet bulb temperature in Atlanta, GA (°F).

## 6.4.1   MINIMUM DISCHARGE PRESSURE

All systems have a certain minimum pressure below which they cannot function correctly. Many factors may contribute to this constraint including pressure drop and vertical head in the piping carrying the compressed vapor as well as hot gas defrost requirements. But by far the most common limitation is the means by which the compressor is cooled. A common approach is liquid re-injection wherein some of the condensed high-pressure refrigerant is diverted to a special port in the compressor where it is expanded and some of the heat of compression is removed. This approach is relatively inexpensive in terms of capital costs, but it comes at a high operational cost. If it uses liquid re-injection, the system is taking a double hit on efficiency. First, some of the high-pressure condensed liquid that required considerable energy to get to that state is expended to cool the compressor and second, this approach generally requires higher discharge pressures to work effectively. The most efficient method to cool screw compressors is the thermosiphon system. These can be expensive to implement, but as we will shortly demonstrate, they are well worth it. In any case, a minimum discharge pressure of 120 psi is usually attainable with reasonable modifications and we shall use that in this case study.

## 6.4.2    CONDENSER FAN WORK AND COMPRESSOR WORK

There is a trade-off between discharge pressure and energy consumed by the fans in the condenser. The higher the pressure, the greater the $\Delta T$ (also known as the approach) that drives the exchange of heat and hence the condenser fans don't have to run as long (or as fast, if they're equipped with variable speed drives). That said, the trade-off isn't equal and it nearly always makes sense to have the condensers work a little harder and the compressor a little less. It would be even better if the system were originally designed with condensers sized such that they can accommodate a lower approach with the same amount of fan work.

## 6.4.3    COMPUTING COMPRESSOR WORK AT VARIOUS WET BULB TEMPERATURES

Now it's simply a case of using our spreadsheet to compute the compressor power required per 100-ton refrigeration capacity at various wet bulb temperatures. Multiply each power by the number of hours that wet bulb is expected to be experienced (Figure 6.3) to get kilowatt-hours consumed, then add up the total. Keep in mind that there will be some wet bulb temperature below which the compressor will have to operate at its minimum discharge pressure of 120 psia.

The first step is to find the wet bulb temperature that corresponds to the minimum discharge pressure, which for this case study is 120 psi. Using Coolprop (or any thermodynamics text) we find that the saturation temperature associated with 120 psi is about 66°F. With a minimal approach of 8°F, we conclude that if the wet bulb temperature is below 58°F, then the discharge pressure will be set to 120 psia. Since 57 is one of the bin boundaries in Figure 6.3, we'll use that as a boundary with little change in accuracy. For all wet bulb temperatures at 57 or lower, the discharge pressure will be set to 120 psia.

Now we go back to the spreadsheet we created in Chapter 5 for the intercooler system and create a new "what if Analysis" table based on the wet bulb temperature. Figure 6.4 shows the beginning of that table.

We start by listing wet bulb temperatures that represent the centers of the bins in Figure 6.3 and end with the temperature that corresponds to the minimum discharge pressure, in this case it's 58°F. Following the directions for a one-dimensional "What-if Analysis" in Excel, we can compute the kW per 100 ton of load the corresponds to each wet bulb temperature range assuming the discharge pressure is just what is required to achieve the minimum approach of 8°F. There is no need to extend the table any lower because the discharge pressure will be maintained at the minimum pressure for all those cases.

Next, we add columns to the table by computing the saturation condenser temperature, which is simply the web bulb temperature plus the approach. From there, the saturation pressure (which is the discharge pressure) is computed using a CoolProp call. Note that the formula used in the table converted the saturation temperature to K and the resulting pressure from Pa to psi. See Figure 6.5.

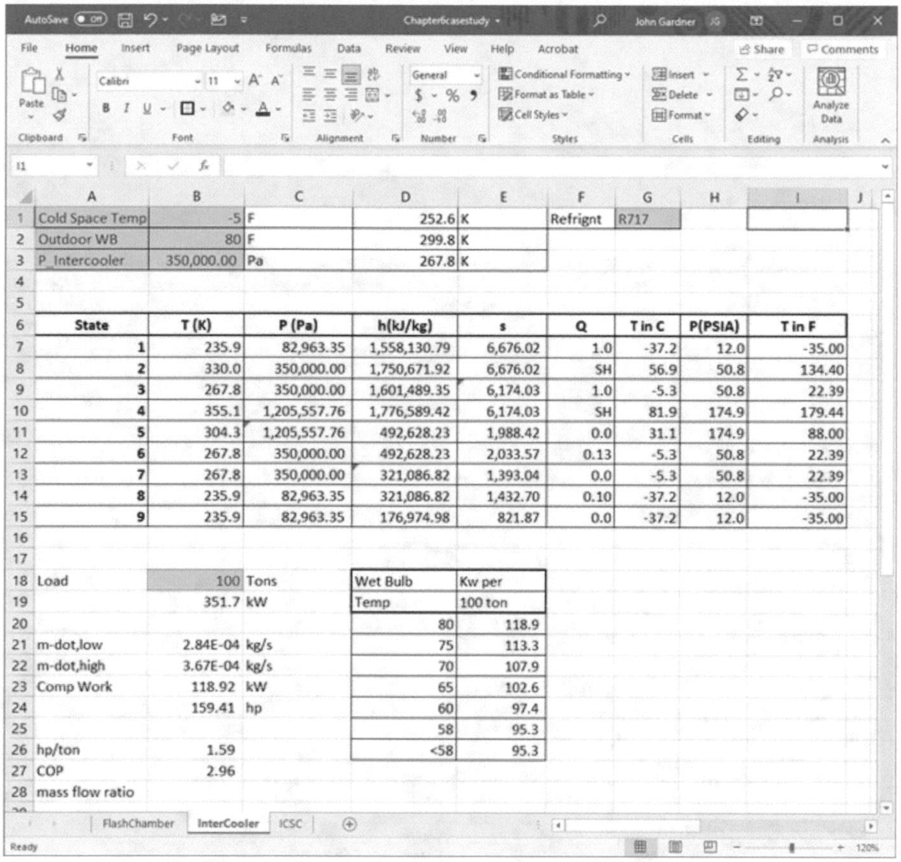

| | A | B | C | D | E | F | G | H | I |
|---|---|---|---|---|---|---|---|---|---|
| 1 | Cold Space Temp | -5 | F | 252.6 | K | Refrignt | R717 | | |
| 2 | Outdoor WB | 80 | F | 299.8 | K | | | | |
| 3 | P_Intercooler | 350,000.00 | Pa | 267.8 | K | | | | |
| 4 | | | | | | | | | |
| 5 | | | | | | | | | |
| 6 | State | T (K) | P (Pa) | h(kJ/kg) | s | Q | T in C | P(PSIA) | T in F |
| 7 | 1 | 235.9 | 82,963.35 | 1,558,130.79 | 6,676.02 | 1.0 | -37.2 | 12.0 | -35.00 |
| 8 | 2 | 330.0 | 350,000.00 | 1,750,671.92 | 6,676.02 | SH | 56.9 | 50.8 | 134.40 |
| 9 | 3 | 267.8 | 350,000.00 | 1,601,489.35 | 6,174.03 | 1.0 | -5.3 | 50.8 | 22.39 |
| 10 | 4 | 355.1 | 1,205,557.76 | 1,776,589.42 | 6,174.03 | SH | 81.9 | 174.9 | 179.44 |
| 11 | 5 | 304.3 | 1,205,557.76 | 492,628.23 | 1,988.42 | 0.0 | 31.1 | 174.9 | 88.00 |
| 12 | 6 | 267.8 | 350,000.00 | 492,628.23 | 2,033.57 | 0.13 | -5.3 | 50.8 | 22.39 |
| 13 | 7 | 267.8 | 350,000.00 | 321,086.82 | 1,393.04 | 0.0 | -5.3 | 50.8 | 22.39 |
| 14 | 8 | 235.9 | 82,963.35 | 321,086.82 | 1,432.70 | 0.10 | -37.2 | 12.0 | -35.00 |
| 15 | 9 | 235.9 | 82,963.35 | 176,974.98 | 821.87 | 0.0 | -37.2 | 12.0 | -35.00 |
| 16 | | | | | | | | | |
| 17 | | | | | | | | | |
| 18 | Load | 100 | Tons | Wet Bulb | Kw per | | | | |
| 19 | | 351.7 | kW | Temp | 100 ton | | | | |
| 20 | | | | 80 | 118.9 | | | | |
| 21 | m-dot,low | 2.84E-04 | kg/s | 75 | 113.3 | | | | |
| 22 | m-dot,high | 3.67E-04 | kg/s | 70 | 107.9 | | | | |
| 23 | Comp Work | 118.92 | kW | 65 | 102.6 | | | | |
| 24 | | 159.41 | hp | 60 | 97.4 | | | | |
| 25 | | | | 58 | 95.3 | | | | |
| 26 | hp/ton | 1.59 | | <58 | 95.3 | | | | |
| 27 | COP | 2.96 | | | | | | | |
| 28 | mass flow ratio | | | | | | | | |

Figure 6.4: Begin developing table using wet bulb temperature as the starting point.

For the next column, refer to the histogram analysis we did in Section 6.3 and illustrated in Figure 6.3. Enter the hours per year associated with each bin of temperatures in the appropriate row. Figure 6.6 shows the results, with the hours highlighted in green shading. Note that for the temperatures at 57°F or lower, we simply add all the remaining bins together.

We finish off the analysis by computing the kWh/year associated with each bin (simply kW times hours/year). We can also account for the difference between the thermodynamic work (which is what we've calculated) and actual electrical usage by assuming a 70% overall conversion efficiency. The costs is then computed by applying the average industrial rate of $0.066 per kWh to the total kWh. For purposes of comparison we compute the corresponding costs for the base case of a constant discharge pressure of 175 psi. See Figure 6.7.

The bottom line of our analysis is that the facility operators can save approximately $15,000/year for every 100-ton refrigeration load. Even if the modifications to achieve this costs $10–$20K, the payoff is very attractive.

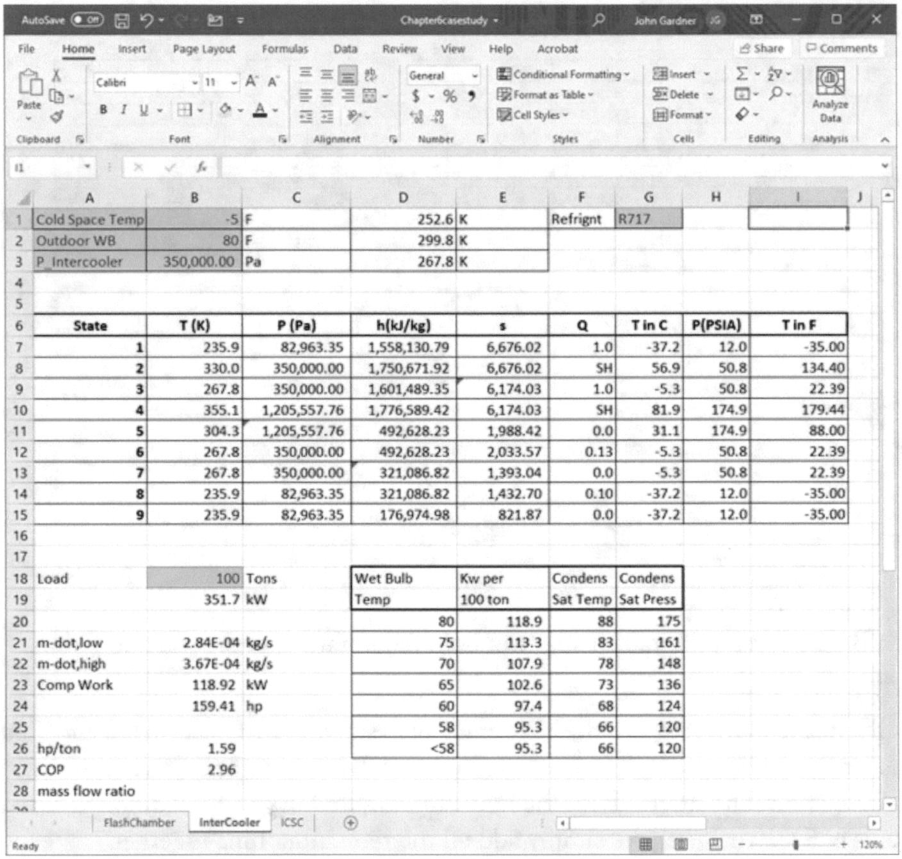

Figure 6.5: Expand table by including saturation temperatures and pressures.

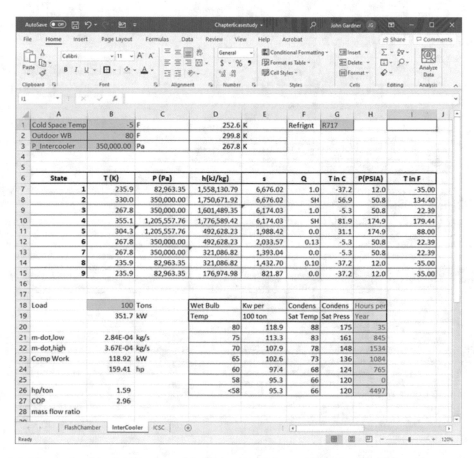

Figure 6.6: Expand table further by adding the hours/year from the TMY bin data.

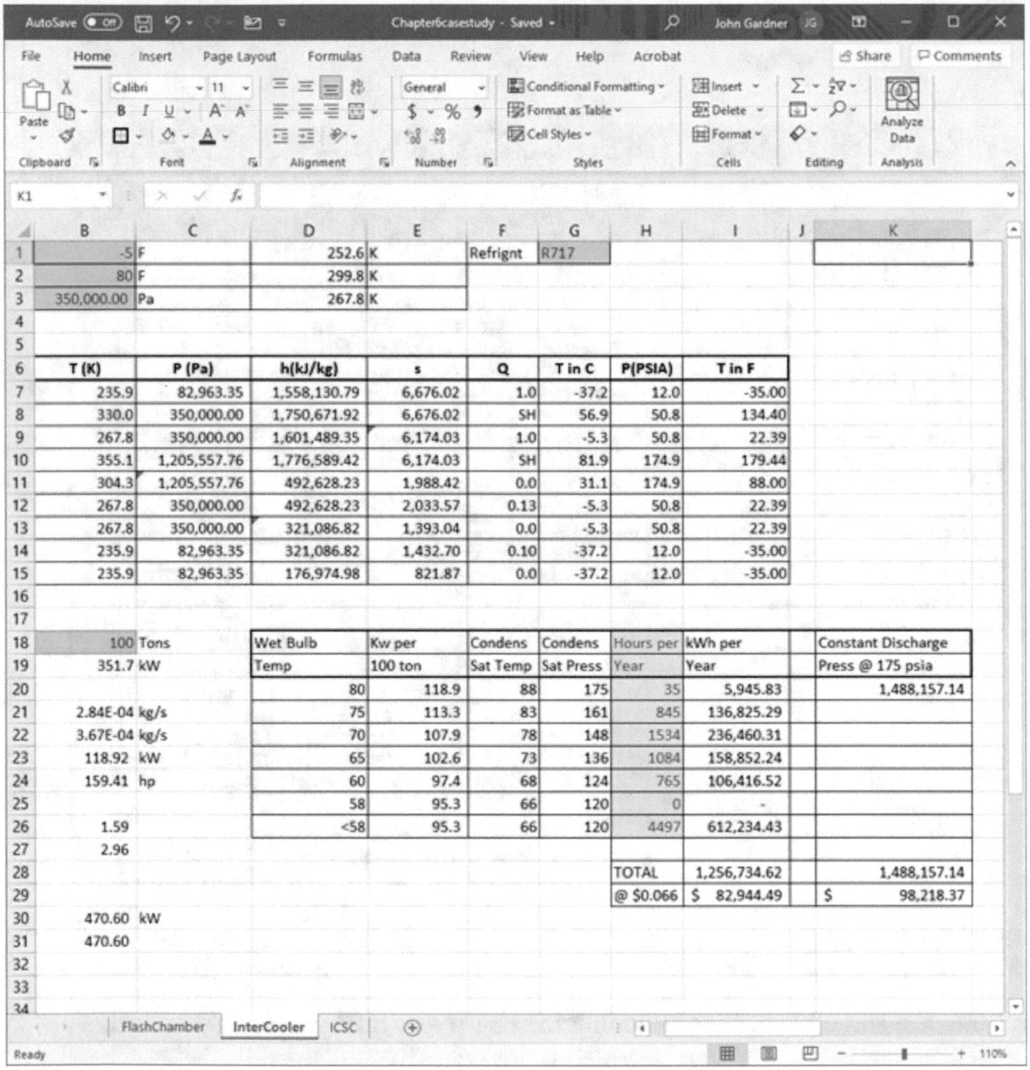

| | B | C | D | E | F | G | H | I | J | K |
|---|---|---|---|---|---|---|---|---|---|---|
| 1 | -5 | F | 252.6 | K | Refrignt | R717 | | | | |
| 2 | 80 | F | 299.8 | K | | | | | | |
| 3 | 350,000.00 | Pa | 267.8 | K | | | | | | |
| 4 | | | | | | | | | | |
| 5 | | | | | | | | | | |
| 6 | T (K) | P (Pa) | h(kJ/kg) | s | Q | T in C | P(PSIA) | T in F | | |
| 7 | 235.9 | 82,963.35 | 1,558,130.79 | 6,676.02 | 1.0 | -37.2 | 12.0 | -35.00 | | |
| 8 | 330.0 | 350,000.00 | 1,750,671.92 | 6,676.02 | SH | 56.9 | 50.8 | 134.40 | | |
| 9 | 267.8 | 350,000.00 | 1,601,489.35 | 6,174.03 | 1.0 | -5.3 | 50.8 | 22.39 | | |
| 10 | 355.1 | 1,205,557.76 | 1,776,589.42 | 6,174.03 | SH | 81.9 | 174.9 | 179.44 | | |
| 11 | 304.3 | 1,205,557.76 | 492,628.23 | 1,988.42 | 0.0 | 31.1 | 174.9 | 88.00 | | |
| 12 | 267.8 | 350,000.00 | 492,628.23 | 2,033.57 | 0.13 | -5.3 | 50.8 | 22.39 | | |
| 13 | 267.8 | 350,000.00 | 321,086.82 | 1,393.04 | 0.0 | -5.3 | 50.8 | 22.39 | | |
| 14 | 235.9 | 82,963.35 | 321,086.82 | 1,432.70 | 0.10 | -37.2 | 12.0 | -35.00 | | |
| 15 | 235.9 | 82,963.35 | 176,974.98 | 821.87 | 0.0 | -37.2 | 12.0 | -35.00 | | |
| 16 | | | | | | | | | | |
| 17 | | | | | | | | | | |
| 18 | 100 | Tons | Wet Bulb | Kw per | Condens | Condens | Hours per | kWh per | | Constant Discharge |
| 19 | 351.7 | kW | Temp | 100 ton | Sat Temp | Sat Press | Year | Year | | Press @ 175 psia |
| 20 | | | 80 | 118.9 | 88 | 175 | 35 | 5,945.83 | | 1,488,157.14 |
| 21 | 2.84E-04 | kg/s | 75 | 113.3 | 83 | 161 | 845 | 136,825.29 | | |
| 22 | 3.67E-04 | kg/s | 70 | 107.9 | 78 | 148 | 1534 | 236,460.31 | | |
| 23 | 118.92 | kW | 65 | 102.6 | 73 | 136 | 1084 | 158,852.24 | | |
| 24 | 159.41 | hp | 60 | 97.4 | 68 | 124 | 765 | 106,416.52 | | |
| 25 | | | 58 | 95.3 | 66 | 120 | 0 | - | | |
| 26 | 1.59 | | <58 | 95.3 | 66 | 120 | 4497 | 612,234.43 | | |
| 27 | 2.96 | | | | | | | | | |
| 28 | | | | | | | TOTAL | 1,256,734.62 | | 1,488,157.14 |
| 29 | | | | | | | @ $0.066 | $  82,944.49 | $ | 98,218.37 |
| 30 | 470.60 | kW | | | | | | | | |
| 31 | 470.60 | | | | | | | | | |
| 32 | | | | | | | | | | |
| 33 | | | | | | | | | | |
| 34 | | | | | | | | | | |

Figure 6.7: Complete the comparison by computing kWh/year at each pressure and add the columns.

## 6.5   SUMMARY

The TMY bin analysis demonstrated in this chapter is a common tool to allow a refrigeration engineer to predict annual performance based on actual weather conditions. Note that a more thorough analysis would include similar models for condenser performance to capture the additional fan work and/or water consumption that would likely result from lower discharge pressures. In the next and final chapter, we will touch on some advanced topics and point the interested reader in the direction of further exploration.

# CHAPTER 7

# Advanced Topics

## 7.1 CARBON DIOXIDE SYSTEMS

Ammonia became the refrigerant of choice for large industrial systems for a number of reasons, but largely due to the fact that ammonia is among the least expansive refrigerants available. Since these systems require large amounts of refrigerant, the cost of the fluid itself is a significant factor. However, ammonia has some serious drawbacks as well, mostly related to the fact that it is extremely toxic and any company using ammonia must comply with extensive regulations to protect worker safety and minimize the chance of environmental impact.

At the same time, there is increasing pressure to eliminate synthetic refrigerants due to their impact on atmospheric ozone and their potential for increasing the planetary greenhouse effect. The original synthetic refrigerants were chlorofluorocarbons (CFCs). CFCs and, to a lesser extent hydrochlorfluorocarbons (HCFCs), have been implicated in the depletion of the earth's ozone layer and have been systematically phased out of use since 1989. This led to the widespread adoption of hydrofluorocarbons (HFCs) through the 1990s. However, HFCs have significant global warming potential and are now under scrutiny by most of the world's governments as global agreements to halt the effects of climate change continue to evolve. As of this writing, the U.S. Environmental Protection Agency is formulating rules to severely limit HFCs starting in 2022.

All this is leading to a resurgence in so-called "natural" (i.e., found in nature) refrigerants, the most common being carbon dioxide. Carbon dioxide ($CO_2$) has many advantages to recommend it as a refrigerant. It is non-corrosive and non-toxic, is one of the least expensive of the refrigerants, has no ozone depletion potential, and is among the lowest global warming potentials (GWP) of any atmospheric gas (GWP = 1). It is also nearly ubiquitously available around the world. On the downside, its thermodynamic properties are such that most common applications require fairly high pressures to achieve the correct evaporator and heat rejection temperatures. To better understand why higher pressures are required for $CO_2$, we can examine the T-S diagram for $CO_2$.

As you'll recall from Chapter 2, you can use the Coolprop package in Python to develop a T-S diagram for various substances. The Python script shown here will create the diagram shown in Figure 7.1.

Let's take for example the baseline case for industrial refrigeration where the cold space is kept at $-5°F$ (252.6 K). The evaporator temperature would likely be considerably lower, say $-20°F$ ( 244K). The T-S diagram shows that the saturation pressure that corresponds to about

**Listing 7.1**   Python code to create a T-S diagram for carbon dioxide

```python
import CoolProp
from CoolProp.Plots import PropertyPlot
#
#  Instantiate the 'PropertyPlot' object for CO2 in SI units
#
plot = PropertyPlot('R744', 'TS', unit_system='SI', tp_limits='ACHP')
#
# Create lines of constant quality
# if you only want to see the dome, use num=2
#
plot.calc_isolines(CoolProp.iQ, num=2)
#
# Create lines of constant pressure (isobars)
#
# plot.calc_isolines(CoolProp.iP, iso_range=[1000000.0,35000000.0],\
#                    num=8,rounding = True)
plot.calc_isolines(CoolProp.iP, \
                   iso_range=[1000000, 2000000,3000000,5000000,8000000,\
                              12000000,20000000,35000000],num=8)
#
# create the plot, modify the iso-lines
#
plot.draw()
plot.isolines.clear()
plot.props[CoolProp.iP]['color'] = 'green'
plot.props[CoolProp.iP]['lw'] = '0.5'
#
#  display and save the plot as a png
#
plot.set_axis_limits([500.0,2500.0,220.0,360.0])
plot.grid()
plot.show()
plot.savefig('CO2-TS.png')
```

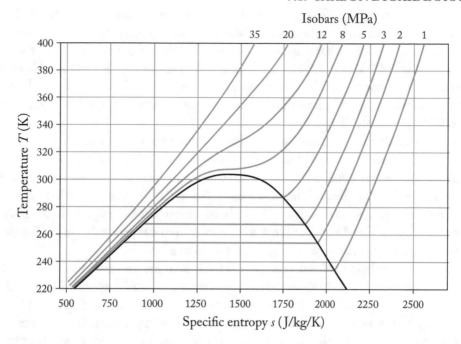

Figure 7.1: T-S diagram for carbon dioxide.

1.5 MPa (218 psia). This is the pressure required for the *low* side of the cycle. Note that this falls within the range of pressures common for the *high* side of refrigeration systems.

The higher pressure is determined by the temperature required to reject the heat. If an evaporative condenser is used, then we need a temperature somewhat higher than about 85°F (302.6 K) which is the highest wet bulb temperature likely to be encountered. Looking at the T-S diagram we can see that if we add any reasonable approach to that temperature, we find ourselves above the saturation dome. The point at the very top of the saturation dome is called the critical point, which for $CO_2$ is at 87.8°F and 1070 psia (304.2 K, 7.38 MPa). The major implication of this fact is that a cycle utilizing $CO_2$ as a refrigerant and rejecting waste heat to the environment must operate above the critical point. These cycles are called transcritical cycles because they span the critical point in their operation.

Heat rejection above the critical point no longer takes place at a constant temperature and there is no condensing happening. For that reason, the heat exchanger utilized is called a gas cooler, not a condenser. Although their appearance may be similar, the operation of a gas cooler is quite different. The isobars in the figure tell the story. The cooling process is still reasonably considered to be isobaric since the pressure drop in the gas cooler will be quite small, resulting only from friction between the gas and the pipe walls and the pressure required to move the gas through the cooler. For the cooler to be effective, the entire process must take place above the design ambient condition. For that reason, the isobar associated with 8 MPa (1160 psia) is likely the lowest discharge pressure reasonable for such a transcritical cycle. Recalling that we model the compression process as isentropic, we find that the point at which has the same specific entropy at 8 MPa as saturated vapor at 244 K is about 205°F (369 K). Of course, the actual discharge temperature will be somewhat less due to the fact that compressors have cooling systems that remove some of the heat of compression, but nonetheless, that's a high temperature, which complicates system design.

This simple example demonstrates another advantage of $CO_2$ systems. To create the full lift between a cold space at −5°F and the environment at 85°F, the compression ratio required is 5.2, which is considerably smaller that the ratio required for other refrigerants. To put this into perspective, for the case studies in Chapters 5 and 6, which utilize the same lift, the single-stage compression ratio required would be nearly 24 whereas the practical design limit for compressors is on the order of 8. This is why systems that serve very cold spaces are multi-stage. The properties of $CO_2$ are such that two stage systems are not always required.

There is one final point about transcritical $CO_2$ systems that is worth noting. Since the discharge temperatures are so high, it makes it easier to find useful applications for the waste heat rather than simply dissipating it to the environment. Space heating, domestic hot water, or various industrial processes could make use of this heat, thus improving the overall efficiency of the larger operation. Finally, it should be noted that a heat exchanger that moves heat from a gas (the compressed $CO_2$) to a liquid (hot water for process applications) is much more compact and efficient when compared to a gas-to-gas heat exchanger such as an evaporative gas cooler.

## 7.2   DYNAMIC SIMULATIONS

In most analysis of industrial refrigeration systems, we assume that the system is running at steady state, although in truth, it rarely does. As material is added and removed from the cold space, as doors open and close and lights turn on and off, the amount of heat to be removed from the cold space (i.e., the refrigeration load) varies. In addition, as we pointed out in Chapter 6, outside weather conditions will impact the ability of the condenser units to reject that heat. A truly comprehensive analysis of these systems requires a dynamic simulation.

A comprehensive treatment of dynamic simulations is well outside the scope of this text and the interested user is referred to any of the numerous textbooks in this field, as exemplified by [KGS07]. What follows is a broad overview of the steps to be taken to develop such a model.

A dynamic simulation starts with a mathematical model of the system. The model starts with an understanding those fundamental variables that define the instantaneous state of the system at any given time. Not surprisingly, these are called the state variables of our system and form the basis of the simulation. The next step is to derive the differential equations that describe how these state variables evolve over time. For a simple example, think of the classic spring/mass/damper system from first year physics. The velocity and displacement of the mass constitute the states and Newton's Second Law ($F = ma$) gives us the differential equation that describes how these states evolve over time.

A simulation is a computer-driven numerical solution of the differential equations that comprise the model. A very common platform for computer simulation of lumped parameter models (i.e., models comprising ordinary differential equations) is Simulink and Matlab.

It may seem surprising, but even though refrigeration systems are very established technology, robust simulations of their operations are not common. The core of the issue is that the condenser and evaporators are heat exchangers with very complex behavior, which becomes more obvious as you consider their operations. Take for example the evaporator. In the case of an over-fed evaporator common to large industrial settings, saturated liquid enters the evaporator and begins absorbing heat due to the air flow being blown across the fins. Immediately the saturated liquid begins to evaporate and the heat transfer is modeled using a boiling liquid heat transfer model [WBB78]. For most evaporators, there exists a location along the way where the refrigerant is 100% vapor, which has very different heat transfer characteristics. What makes this particular case so complex is that the point at which the refrigerant changes from a saturated mixture, to pure vapor changes based on the conditions. Therefore, the location of that boundary must be identified as a state of the system and the differential equation that describes how that changes must be derived.

In recent years, there has been renewed interest in dynamic modeling of refrigeration systems and low-order dynamic models of condensers and evaporators have been developed and validated. The foundational work done by Asada [HLA97] was expanded by Allyene and others [MA08][LA10] and there is now a significant body of work that the properly motivated analyst could use to develop appropriate simulations of their own.

## 7.3    OPTIMIZATION AND DIGITAL TWINS

Refrigeration systems in general are challenging to design and operate due in part to the complex nature of the heat transfer that takes place in the evaporator and condenser. The complexity is compounded with the addition of a booster compressor, intercooler, and evaporative condenser. The designer's task of finding the right components to create a reliable system that will meet the customer's needs is a daunting task that relies heavily on manufacturer's performance charts, rules of thumb, and experience. Similarly, finding the best combination of operational parameters such as intercooler pressure, discharge pressure, and condenser set points is an equally difficult task. The applications of thermodynamic modeling as introduced in this text to the field of dynamic system modeling and simulation opens the door to the use of powerful optimization tools that can lead to control protocols that will save significant energy without sacrificing performance. A robust and validated whole-system simulations specific to a given application is commonly known as a "digital twin." Currently, this level of system description is rare, if available at all, but as modeling expertise expands and as the drive for more efficiency continues, digital twin technology will expand into the industrial refrigeration field in the near future.

## 7.4    NEXT STEPS

In this brief monograph, we have just scratched the surface of bringing modern computational tools to the area of industrial refrigeration systems. But these tools are the essential starting point in any such endeavors. Possible applications span the gamut from advanced computer-aided design tools to digital twin simulations. For these advanced applications, the flexible coding and extensive user bases of Matlab and Python would be extremely useful.

# APPENDIX A

# Matlab Scripts

**Listing A.1** Matlab script to analyze the single-stage configuration *(Continues.)*

```
%  Script to compare various ammonia cycles
%  Set up the baseline parameters for our case
%  Case 1: Single Stage, no intercooling
REFR = 'R717'; % Ammonia
ColdTemp = -5.0; % Temperature of cold space in F
OutsideWB = 80.0; % Worst Case WB in F
%
EvapSatTemp = ColdTemp - 30.;
CondensSatTemp = OutsideWB + 15;
%
Qlow = 100.0; % Tons Refrigeration
% convert to SI
EvapSatTempSI = (EvapSatTemp-32)/1.8+273.15; % K
CondensSatTempSI = (CondensSatTemp-32)/1.8+273.15; %K
QlowSI = Qlow*3.517; %kW
%  Compute State Properties
P1 = py.CoolProp.CoolProp.PropsSI('P','T',EvapSatTempSI,'Q',1,REFR);
P3 = py.CoolProp.CoolProp.PropsSI('P','T',CondensSatTempSI,'Q',0,REFR);
P2 = P3;
P4 = P1;
%
s1 = py.CoolProp.CoolProp.PropsSI('S','T',EvapSatTempSI,'Q',1,REFR);
s2 = s1; %assume isentropic compression
%
h1 = py.CoolProp.CoolProp.PropsSI('H','T',EvapSatTempSI,'Q',1,REFR);
h2 = py.CoolProp.CoolProp.PropsSI('H','P',P2,'S',s2,REFR);
h3 = py.CoolProp.CoolProp.PropsSI('H','T',CondensSatTempSI,'Q',0,REFR);
h4 = h3;
```

**Listing A.1**   *(Continued.)* Matlab script to analyze the single-stage configuration

```
%
T2 = py.CoolProp.CoolProp.PropsSI('T','P',P2,'S',s2,REFR);
% Compute Energy Values
qlow = h1 - h4;
mdot=QlowSI/qlow;
Wdot = mdot*(h2-h1);
COP = QlowSI/Wdot;
```

**Listing A.2**  Matlab script to analyze the flash chamber configuration *(Continues.)*

```matlab
%  Script to compare various ammonia cycles
%  Set up the baseline parameters for our case
%  Case 2: Two-stage, Flash Chamber
REFR = 'R717'; % Ammonia
ColdTemp = -5.0;   % Temperature of cold space in F
OutsideWB = 80.0; % Worst Case WB in F
EvapSatTemp = ColdTemp - 30.;
CondensSatTemp = OutsideWB + 15;
Qlow = 100.0; % Tons Refrigeration
Pintercooler = 380000.0; % Pa, from chapter 5 case study
% convert to SI
EvapSatTempSI = (EvapSatTemp-32)/1.8+273.15; % K
CondensSatTempSI = (CondensSatTemp-32)/1.8+273.15; %K
QlowSI = Qlow*3.517; %kW
%  Compute State Properties
P1 = py.CoolProp.CoolProp.PropsSI('P','T',EvapSatTempSI,'Q',1,REFR);
P5 = py.CoolProp.CoolProp.PropsSI('P','T',CondensSatTempSI,'Q',0,REFR);
P8 = P1; P4 = P5;
P3 = Pintercooler;
P9 = P3; P6 = P3; P7 = P3; P2 = P3;
%
s1 = py.CoolProp.CoolProp.PropsSI('S','T',EvapSatTempSI,'Q',1,REFR);
s2 = s1;   %assume isentropic compression
%
h1 = py.CoolProp.CoolProp.PropsSI('H','T',EvapSatTempSI,'Q',1,REFR);
h2 = py.CoolProp.CoolProp.PropsSI('H','P',P2,'S',s2,REFR);
h5 = py.CoolProp.CoolProp.PropsSI('H','T',CondensSatTempSI,'Q',0,REFR);
h6 = h5;
h7 = py.CoolProp.CoolProp.PropsSI('H','P',P7,'Q',0,REFR);
h8 = h7;
h9 = py.CoolProp.CoolProp.PropsSI('H','P',P9,'Q',1,REFR);
```

**Listing A.2**   *(Continued.)* Matlab script to analyze the flash chamber configuration

```
%
Q6 = py.CoolProp.CoolProp.PropsSI('Q','P',P6,'H',h6,REFR);
h3 = Q6*h9+(1-Q6)*h2;
%
s3 = py.CoolProp.CoolProp.PropsSI('S','P',P3,'H',h3,REFR);
s4 = s3;
h4 = py.CoolProp.CoolProp.PropsSI('H','P',P4,'S',s4,REFR);
T4 = py.CoolProp.CoolProp.PropsSI('T','P',P4,'S',s4,REFR);
% Compute Energy Values
qlow = h1 - h8;
mdotlow=QlowSI/qlow;
mdothigh = mdotlow/(1-Q6);
Wdot = mdotlow*(h2-h1)+mdothigh*(h4-h3); % sum of both compressors
COP = QlowSI/Wdot;
```

**Listing A.3**   Matlab script to analyze the intercooler configuration *(Continues.)*

```matlab
%  Script to compare various ammonia cycles
%  Set up the baseline parameters for our case
%  Case 3: Two-stage, Intercooler
REFR = 'R717'; % Ammonia
ColdTemp = -5.0;   % Temperature of cold space in F
OutsideWB = 80.0; % Worst Case WB in F
EvapSatTemp = ColdTemp - 30.;
CondensSatTemp = OutsideWB + 15;
Qlow = 100.0; % Tons Refrigeration
Pintercooler = 370000.0; % Pa, from chapter 5 case study
% convert to SI
EvapSatTempSI = (EvapSatTemp-32)/1.8+273.15; % K
CondensSatTempSI = (CondensSatTemp-32)/1.8+273.15; %K
QlowSI = Qlow*3.517; %kW
%  Compute State Properties
P1 = py.CoolProp.CoolProp.PropsSI('P','T',EvapSatTempSI,'Q',1,REFR);
P5 = py.CoolProp.CoolProp.PropsSI('P','T',CondensSatTempSI,'Q',0,REFR);
P8 = P1; P4 = P5;
P3 = Pintercooler;
P6 = P3; P7 = P3; P2 = P3;
%
s1 = py.CoolProp.CoolProp.PropsSI('S','T',EvapSatTempSI,'Q',1,REFR);
s2 = s1;   %assume isentropic compression
h1 = py.CoolProp.CoolProp.PropsSI('H','T',EvapSatTempSI,'Q',1,REFR);
h2 = py.CoolProp.CoolProp.PropsSI('H','P',P2,'S',s2,REFR);
h3 = py.CoolProp.CoolProp.PropsSI('H','P',P3,'Q',1,REFR);
h5 = py.CoolProp.CoolProp.PropsSI('H','T',CondensSatTempSI,'Q',0,REFR);
h6 = h5;
h7 = py.CoolProp.CoolProp.PropsSI('H','P',P7,'Q',0,REFR);
h8 = h7;
```

**Listing A.3**   *(Continued.)* Matlab script to analyze the intercooler configuration

```
%
s3 = py.CoolProp.CoolProp.PropsSI('S','P',P3,'H',h3,REFR);
s4 = s3;
%
h4 = py.CoolProp.CoolProp.PropsSI('H','P',P4,'S',s4,REFR);
T4 = py.CoolProp.CoolProp.PropsSI('T','P',P4,'S',s4,REFR);
% Compute Energy Values
x = (h7-h2)/(h6+h7-h2-h3);
qlow = h1 - h8;
mdotlow=QlowSI/qlow;
mdothigh = x/(1-x)*mdotlow;
Wdot = mdotlow*(h2-h1)+mdothigh*(h4-h3); % sum of both compressors
COP = QlowSI/Wdot;
```

**Listing A.4** Matlab script to analyze the intercooler + subcooler configuration *(Continues.)*

```
%  Script to compare various ammonia cycles
%  Set up the baseline parameters for our case
%  Case 4: Two-stage, Intercooler + SubCooler
REFR = 'R717'; % Ammonia
ColdTemp = -5.0;  % Temperature of cold space in F
OutsideWB = 80.0; % Worst Case WB in F
EvapSatTemp = ColdTemp - 30.;
CondensSatTemp = OutsideWB + 15;
Qlow = 100.0; % Tons Refrigeration
Pintercooler = 380000.0; % Pa, from chapter 5 case study
DeltT = 10;  % Subcooling, F
% convert to SI
EvapSatTempSI = (EvapSatTemp-32)/1.8+273.15; % K
CondensSatTempSI = (CondensSatTemp-32)/1.8+273.15; %K
QlowSI = Qlow*3.517; %kW
DeltaTSI = 10/1.8;
%  Compute State Properties
P1 = py.CoolProp.CoolProp.PropsSI('P','T',EvapSatTempSI,'Q',1,REFR);
P5 = py.CoolProp.CoolProp.PropsSI('P','T',CondensSatTempSI,'Q',0,REFR);
P8 = P1; P4 = P5; P7 = P5;
P3 = Pintercooler;
P6 = P3; P2 = P3;
T5 = py.CoolProp.CoolProp.PropsSI('T','P',P5,'Q',0,REFR);
T7 = T5-DeltaTSI;
s1 = py.CoolProp.CoolProp.PropsSI('S','T',EvapSatTempSI,'Q',1,REFR);
s2 = s1;  %assume isentropic compression
h1 = py.CoolProp.CoolProp.PropsSI('H','T',EvapSatTempSI,'Q',1,REFR);
h2 = py.CoolProp.CoolProp.PropsSI('H','P',P2,'S',s2,REFR);
h3 = py.CoolProp.CoolProp.PropsSI('H','P',P3,'Q',1,REFR);
h5 = py.CoolProp.CoolProp.PropsSI('H','T',CondensSatTempSI,'Q',0,REFR);
h6 = h5;
h7 = py.CoolProp.CoolProp.PropsSI('H','P',P7,'T',T7,REFR);
h8 = h7;
```

**Listing A.4** *(Continued.)* Matlab script to analyze the intercooler + subcooler configuration

```matlab
s3 = py.CoolProp.CoolProp.PropsSI('S','P',P3,'H',h3,REFR);
s4 = s3;
h4 = py.CoolProp.CoolProp.PropsSI('H','P',P4,'S',s4,REFR);
T4 = py.CoolProp.CoolProp.PropsSI('T','P',P4,'S',s4,REFR);
% Compute Energy Values
x = (h7+h3-h5-h2)/(h6+h7-h2-h5);
qlow = h1 - h8;
mdotlow=QlowSI/qlow;
mdothigh = 1/(1-x)*mdotlow;
Wdot = mdotlow*(h2-h1)+mdothigh*(h4-h3); % sum of both compressors
COP = QlowSI/Wdot;
```

# Bibliography

[WBB78]   GL Wedekind, BL Bhatt, and BT Beck. "A system mean void fraction model for predicting various transient phenomena associated with two-phase evaporating and condensing flows". In: *International Journal of Multiphase Flow* 4.1 (1978), pp. 97–114.

[HLA97]   Xiang-Dong He, Sheng Liu, and Haruhiko H. Asada. "Modeling of Vapor Compression Cycles for Multivariable Feedback Control of HVAC Systems". In: *Journal of Dynamic Systems, Measurement, and Control* 119.2 (June 1997), pp. 183–191. ISSN: 0022-0434. DOI: 10.1115/1.2801231. URL: http://dx.doi.org/10.1115/1.2801231.

[WMB04]   Marcus Wilcox, Rob Morton, and Dan Brown. *Industrial Refrigeration Best Practices Guide*. Cascade Energy, 2004.

[KGS07]   Bohdan T. Kulakowski, John F. Gardner, and J. Lowen Shearer. *Dynamic modeling and control of engineering systems*. 3rd ed. Section: xii, 486 pages : illustrations ; 27 cm. Cambridge: Cambridge University Press, 2007. ISBN: 978-0-521-86435-0 978-0-511-28942-2. URL: http://catdir.loc.gov/catdir/enhancements/fy0668/2006031544-t.html (visited on 06/03/2021).

[MA08]   Thomas L. McKinley and Andrew G. Alleyne. "An advanced nonlinear switched heat exchanger model for vapor compression cycles using the moving-boundary method". In: *International Journal of Refrigeration* 31.7 (Nov. 2008), pp. 1253–1264. ISSN: 0140-7007. DOI: 10.1016/j.ijrefrig.2008.01.012. URL: http://www.sciencedirect.com/science/article/pii/S0140700708000327.

[LA10]   Bin Li and Andrew G. Alleyne. "A dynamic model of a vapor compression cycle with shut-down and start-up operations". In: *International Journal of Refrigeration* 33.3 (May 2010), pp. 538–552. ISSN: 0140-7007. DOI: 10.1016/j.ijrefrig.2009.09.011. URL: http://www.sciencedirect.com/science/article/pii/S0140700709002126.

[Joh13]   Sherena Johnson. *REFPROP*. en. text. Last Modified: 2020-11-16T17:06-05:00. Apr. 2013. URL: https://www.nist.gov/srd/refprop (visited on 05/12/2021).

[Bel+14]   Ian H. Bell et al. "Pure and Pseudo-pure Fluid Thermophysical Property Evaluation and the Open-Source Thermophysical Property Library CoolProp". In: *Industrial & Engineering Chemistry Research* 53.6 (2014), pp. 2498–2508. DOI: 10.1021/ie4033999. eprint: http://pubs.acs.org/doi/pdf/10.1021/ie4033999. URL: http://pubs.acs.org/doi/abs/10.1021/ie4033999.

[Mor14]   Michael J. Moran, ed. *Fundamentals of engineering thermodynamics*. 8th ed. OCLC: ocn879865441. Hoboken, NJ: Wiley, 2014. ISBN: 978-1-118-41293-0 978-1-118-82044-5.

# Author's Biography

## JOHN F. GARDNER

**John Gardner** is professor emeritus of mechanical and biomedical engineering at Boise State University where he held many appointments including department chair (2001–2007, 2020–2021), Associate Vice President (2007–2010), and Director of the CAES Energy Efficiency Research Institute (2010–2020). Prior to that, he was on the faculty of the Mechanical and Nuclear Engineering Department at Penn State (1987–2000). Gardner earned his Bachelor's degree in mechanical engineering at Cleveland State University and his Master's and Ph.D. in mechanical engineering at The Ohio State University. He is widely published in the areas of robotics, modeling and control of dynamic systems, biomedical engineering, and engineering education. He is a licensed professional engineer in the state of Idaho and a Fellow of the American Society of Mechanical Engineers.

Printed in the United States
by Baker & Taylor Publisher Services